TEAM HUMAN

Douglas Rushkoff

W. W. NORTON & COMPANY

Independent Publishers Since 1923

NEW YORK   LONDON

For information about permission to reproduce selections from this book,
write to Permissions, W. W. Norton & Company, Inc.,
500 Fifth Avenue, New York, NY 10110

For information about special discounts for bulk purchases, please contact
W. W. Norton Special Sales at specialsales@wwnorton.com or 800-233-4830

Manufacturing by Worzalla Publishing Company
Book design by Lovedog Studio
Production manager: Beth Steidle

ISBN 978-0-393-65169-0

W. W. Norton & Company, Inc., 500 Fifth Avenue, New York, N.Y. 10110
www.wwnorton.com

W. W. Norton & Company Ltd., 15 Carlisle Street, London W1D 3BS

1 2 3 4 5 6 7 8 9 0

CONTENTS

Find the Others . . .

TEAM
HUMAN

1.

Autonomous technologies, runaway markets, and weaponized media seem to have overturned civil society, paralyzing our ability to think constructively, connect meaningfully, or act purposefully. It feels as if civilization itself were on the brink, and that we lack the collective willpower and coordination necessary to address issues of vital importance to the very survival of our species.

It doesn't have to be this way.

2.

Everyone is asking how we got here, as if this were a random slide toward collective incoherence and disempowerment. It is not. There's a reason for our current predicament: an antihuman agenda embedded in our technology, our markets, and our major cultural institutions, from education and religion to civics and media. It has turned them from forces for human connection and expression into ones of isolation and repression.

By unearthing this agenda, we render ourselves capable of transcending its paralyzing effects, reconnecting to one another, and remaking society toward human ends rather than the end of humans.

3.

The first step toward reversing our predicament is to recognize that being human is a team sport. We cannot be fully human alone. Anything that brings us together fosters our humanity.

Likewise, anything that separates us makes us less human, and less able to exercise our individual or collective will.

We use our social connections to orient ourselves, to ensure mutual survival, and to derive meaning and purpose. This is not merely a quaint notion but our biological legacy. People who are disconnected from the organizations or communities they serve often wither without them.

We sometimes connect with one another in order to achieve some common goal, such as finding food or evading prey. But we also commune and communicate for their own sake—because we gain strength, pleasure, and purpose as we develop rapport. Are you there? Yes, I hear you.

You are not alone.

4.

We have amplified and extended our natural ability to connect by inventing various forms of media. Even a one-way medium, such as a book, creates a new intimacy as it lets us see the world through another person's eyes. Television enables us to bear witness to what is happening to people across the globe, and to do so en masse. On TV we watched together, simultaneously, events from the moon landing to the felling of the Berlin Wall, and experienced our collective humanity as never before.

Likewise, the internet connects us more deliberately and, in some ways, more reassuringly than any medium before it. With its development, the tyranny of top-down broadcast media seemed to be broken by the peer-to-peer connections and free expressions of every human node on the network. The net turned media back into a collective, participatory, and social landscape.

But, as seemingly happens to each and every new medium, the net went from being a social platform to an isolating one.

Instead of forging new relationships between people, our digital technologies came to replace them with something else.

We live with a bounty of communications technologies at our disposal. Our culture is composed more of mediated experiences than of directly lived ones. Yet we are also more alone and atomized than ever before. Our most advanced technologies are not enhancing our connectivity, but thwarting it. They are replacing and devaluing our humanity, and—in many different ways—undermining our respect for one another and ourselves. Sadly, this has been by design. But that's also why it can be reversed.

5.

We are embedding some very old and disparaging notions about human beings and their place in the natural order into our future technological infrastructure. Engineers at our leading tech firms and universities tend to see human beings as the problem and technology as the solution.

When they are not developing interfaces to control us, they are building intelligences to replace us. Any of these technologies could be steered toward extending our human capabilities and collective power. Instead, they are deployed in concert with the demands of a marketplace, political sphere, and power structure that depend on human isolation and predictability in order to operate.

Social control is based on thwarting social contact and exploiting the resulting disorientation and despair. Human beings evolved by gaining the capacity to forge greater numbers of social connections. The development of our brain, language, text, electronic media, and digital networks were all driven by our need for higher levels of social organization. The net—just the latest of these advances—challenges us with the possibility that thinking and memory may not be personal at all, but group activities. But this potential has been overshadowed by deep suspicion of how human beings might behave as an empowered collective, as well as a growing awareness that socially fulfilled people need less money, experience less shame, behave less predictably, and act more autonomously.

Thinking, feeling, connected people undermine the institutions that would control them. They always have. That's why new mechanisms for forging bonds and cooperation between people are almost inevitably turned against those ends. Language that could inform is instead used to lie. Money that could promote trade is instead hoarded by the wealthy. Education that could expand workers' minds is instead used to make them more efficient human resources.

All along the way, cynical views of humans as a mindless mob, incapable of behaving intelligently and peacefully, are used to justify keeping us apart and denying us roles as autonomous actors in any of these areas of life. Our institutions and technologies aren't designed to extend our human nature, but to mitigate or repress it.

Once our humanity is seen as a liability instead of a strength, the corresponding cultural drive and spiritual quest is to transcend our personhood: a journey out of body, away from our humanness, beyond matter, and into whatever substrate—be it ether, electrical wavelengths, virtual reality, or AI—we fetishize at that moment.

6.

Digital networks are just the latest media to go from promoting social bonds to destroying them—from fostering humanity to supplanting it. Our current shift may be more profound and permanent, however, because this time we are empowering our antihuman technologies with the ability to retool themselves. Our smart devices iterate and evolve faster than our biology can.

We are also tying our markets and our security to the continued growth and expanding capabilities of our machines. This is self-defeating. We are increasingly depending on technologies built with the presumption of human inferiority and expendability.

But the unprecedented speed of this latest reversal from social extension to social annihilation also offers us an opportunity to understand the process by which it happens. Once we do, we will recognize how it has occurred in myriad ways throughout history—from agriculture and education to currency and democracy.

We humans—in a single generation—are experiencing a turn of the cycle in real time. This is our chance. We can choose not to adapt to it any longer, but to oppose it.

7.

It's time we reassert the human agenda. And we must do so together—not as the individual players we have been led to imagine ourselves to be, but as the team we actually are.

Team Human.

8.

Nature is a collaborative act. If humans are the most evolved species, it is only because we have developed the most advanced ways of working and playing together.

We've been conditioned to believe in the myth that evolution is about competition: the survival of the fittest. In this view, each creature struggles against all the others for scarce resources. Only the strongest ones survive to pass on their superior genes, while the weak ones live to lose and die out.

But evolution is every bit as much about cooperation as competition. Our very cells are the result of an alliance billions of years ago between mitochondria and their hosts. Individuals and species flourish by evolving ways of supporting mutual survival. A bird develops a beak which lets it feed on some part of a plant that other birds can't reach. This introduces diversity into the population's diet, reducing the strain on a particular food supply and leading to more for all. What of the poor plant, you ask? The birds, much like bees, are helping the plant by spreading its seeds after eating its fruit.

Survival of the fittest is a convenient way to justify the cutthroat ethos of a competitive marketplace, political landscape, and culture. But this perspective misconstrues the theories of Darwin as well as his successors. By viewing evolution though a strictly competitive lens, we miss the bigger story of our own social development and have trouble understanding humanity as one big, interconnected team.

The most successful of biology's creatures coexist in mutually beneficial ecosystems. It's hard for us to recognize such widespread cooperation. We tend to look at life forms as isolated from one another: a tree is a tree and a cow is a cow. But a tree is not a singular tree at all; it is the tip of a forest. Pull back far enough to see the whole, and one tree's struggle for survival merges with the more relevant story of its role in sustaining the larger system.

We also tend to miss nature's interconnections because they happen subtly, beneath the surface. We can't readily see or hear the way trees communicate. For instance, there's an invisible landscape of mushrooms and other fungi connecting the root systems of trees in a healthy forest. The underground network allows the trees to interact with one another and even exchange resources. In the summer, shorter evergreens are shaded by the canopies of taller trees. Incapable of reaching the light and photosynthesizing, they call through the fungus for the sun-drenched nutrients they need. The taller trees have plenty to spare, and send it to their shaded peers. The taller trees lose their leaves in the winter and themselves become incapable of photosynthesizing. At that point, the evergreens, now exposed to the sun, send their extra nutrients to their leafless community members. For their part, the underground fungi charge a small service fee, taking the nutrients they need in return for facilitating the exchange.

So the story we are taught in school about how trees of the forest compete to reach the sunlight isn't really true. They collaborate to reach the sunlight, by varying their strategies and sharing the fruits of their labor.

Trees protect one another as well. When the leaves of acacia trees come in contact with the saliva of a giraffe, they release a warning chemical into the air, triggering nearby acacias to release repellents specific to giraffes. Evolution has raised them to behave as if they were part of the same, self-preserving being.

9.

Animals cooperate as well. Their mutually beneficial behaviors are not an exception to natural selection, but the rule.

Darwin observed how wild cattle could tolerate only a brief separation from their herd, and slavishly followed their leaders. "Individualists" who challenged the leader's authority or wandered away from the group were picked off by hungry lions. Darwin generalized that social bonding was a "product of selection." In other words, teamwork was a better strategy for everyone's survival than competition.

Darwin saw what he believed were the origins of human moral capabilities in the cooperative behavior of animals. He marveled at how species from pelicans to wolves have learned to hunt in groups and share the bounty, and how baboons expose insect nests by cooperating to lift heavy rocks.

Even when they are competing, many animals employ social strategies to avoid life-threatening conflicts over food or territory. Like break-dancers challenging one another in a ritualized battle, the combatants assume threatening poses or inflate their chests. They calculate their relative probability of winning an all-out conflict and then choose a winner without actually fighting.

The virtual combat benefits not just the one who would be killed, but also the victor, who could still be injured. The loser is free to go look for something else to eat, rather than wasting time or losing limbs in a futile fight.

10.

Evolution may have less to do with rising above one's peers than learning to get along with more of them.

We used to believe that human beings developed larger brains than chimpanzees in order to do better spatial mapping of our environment or to make more advanced tools and weapons. From a simplistic survival-of-the-fittest perspective, this makes sense. Primates with better tools and mental maps would hunt and fight better, too. But it turns out there are only slight genetic variations between hominids and chimpanzees, and they relate almost exclusively to the number of neurons that our brains are allowed to make. It's not a qualitative difference but a quantitative one. The most direct benefit of more neurons and connections in our brains is an increase in the size of the social networks we can form. Complicated brains make for more complex societies.

Think of it this way: a quarterback, point guard, or midfielder, no matter their skills, is only as valuable as their ability to coordinate with the other players; a great athlete is one who can predict the movements of the most players at the same time. Similarly, developing primates were held back less by their size or skills than by their social intelligence. Bigger groups of primates survived better, but required an increase in their ability to remember everyone, manage relationships, and coordinate activities. Developing bigger brains allowed human beings to maintain a whopping 150 stable relationships at a time.

The more advanced the primate, the bigger its social groups. That's the easiest and most accurate way to understand evolution's trajectory, and the relationship of humans to it. Even if we don't agree that social organization is evolution's master plan, we must accept that it is—at the very least—a large part of what makes humans human.

11.

Human social cohesion is supported by subtle biological processes and feedback mechanisms. Like trees that communicate through their root systems, human beings have developed elaborate mechanisms to connect and share with one another.

Our nervous systems learned to treat our social connections as existentially important—life or death. Threats to our relationships are processed by the same part of the brain that processes physical pain. Social losses, such as the death of a loved one, divorce, or expulsion from a social group, are experienced as acutely as a broken leg.

Managing social relationships also required humans to develop what anthropologists call a "theory of mind"—the ability to understand and identify with the thinking and motivations of other people. From an evolutionary perspective, the concept of self came after our ability to evaluate and remember the intentions and tactics of others. Unlike the relatively recent cultural changes that encouraged ideas of personal identity or achievement, our social adaptations occurred over hundreds of thousands of years of biological evolution. Enduring social bonds increase a group's ability to work together, as well as its chances for procreation. Our eyes, brains, skin, and breathing are all optimized to enhance our connection to other people.

Prosocial behaviors such as simple imitation—what's known as mimesis—make people feel more accepted and included, which sustains a group's cohesion over time. In one experiment, people who were subtly imitated by a group produced less stress hormone than those who were not imitated. Our bodies are adapted to seek and enjoy being mimicked. When human beings are engaged in mimesis, they learn from one another and advance their community's skill set.

The physical cues we use to establish rapport are preverbal. We used them to bond before we ever learned to speak—both as babies and as early humans many millennia ago. We flash our eyebrows when we want someone to pay attention to us. We pace someone else's breathing when we want them to know we empathize. The pupils of our eyes dilate when we feel open to what another person is offering. In turn, when we see someone breathing with us, their eyes opening to accept us, their head subtly nodding, we feel we are being understood and accepted. Our mirror neurons activate, releasing oxytocin—the bonding hormone—into our bloodstream.

Human beings connect so easily, it's as if we share the same brains. Limbic consonance, as it's called, is our ability to attune to one another's emotional states. The brain states of mothers and their babies mirror each other; you can see this in an MRI scan. Limbic consonance is the little-known process through which the mood of a room changes when a happy or nervous person walks in, or the way a person listening to a story acquires the same brain state as the storyteller. Multiple nervous systems sync and respond together, as if they were one thing. We long for such consonance, as well as the happy hormones and neural regulation that come with it. It's why our kids want to sleep with us—their nervous systems learn how to sleep and wake by mirroring ours. It's why television comedies have laugh tracks—so that we are coaxed to imitate the laughter of an audience of peers watching along. We naturally try to resonate with the brain state of the crowd.

These painstakingly evolved, real-world physical and chemical processes are what enable and reinforce our social connection and coherence, and form the foundations for the societies that we eventually built.

12.

Thanks to organic social mechanisms, humans became capable of pair bonding, food sharing, and even collective childcare.

Our survivability increased as we learned how to orchestrate simple divisions of labor, and trusted one another enough to carry them out.

The more spectacular achievement was not the division of labor but the development of group sharing. This distinguished true humans from other hominids: we waited to eat until we got the bounty back home. Humans are defined not by our superior hunting ability so much as by our capacity to communicate, trust, and share.

Biologists and economists alike have long rejected social or moral justifications for this sort of behavior. They chalk it up instead to what they call "reciprocal altruism." One person does a nice thing for another person in the hope of getting something back in the future. You take a risk to rescue someone else's child from a dangerous predator because you trust the other parent to do the same for your kid. In this view, people aren't so nice at all; they're just acting on their own behalf in a more complicated way.

But contemporary research strongly supports more generous motives in altruism, which have nothing to do with self-interest. Early humans had a strong disposition to cooperate with one another, at great personal cost, even when there could be no expectation of payback in the future. Members of a group who violated the norms of cooperation were punished. Solidarity and community were prized in their own right.

Evolution's crowning achievement, in this respect, was the emergence of spoken language. It was a dangerous adaptation that involved crossing the airway with the foodway, making us vulnerable to choking. But it also gave us the ability to modify the sounds that came from our vocal folds and make the variety of mouth noises required for language.

While language may have been driven by the need for larger,

more complicated social structures, think of the immense collaborative act that developing a language required from its speakers. That multigenerational exercise alone would change the fabric of society and its faith in a cooperative enterprise.

13.

Language changed everything. Once people acquired speech, cultural development and social cohesion no longer depended on increasing our brain size. Evolution shifted from a purely biological process to a social one. With language, humans gained the ability to learn from one another's experiences. The quest for knowledge began.

Other animals, such as apes, learn by doing. Episodic learning, as it's called, means figuring things out oneself, through trial and error. Fire is hot. If you can remember what happened last time you touched it, you don't touch it again. Even simpler creatures store the equivalent of learning as instincts or natural behaviors, but they are procedural and automatic. Humans, on the other hand, can learn by imitating one another or, better, representing their experiences to one another through language. This is big, and may give us the clearest way of understanding what it means to be human.

The difference between plants, animals, and humans comes down to what each life form can store, leverage, or—as this concept has been named—"bind." Plants can bind energy. They transform sunlight into biological energy. By spreading their leaves, they harvest ultraviolet rays and turn them into energy that they (and the animals that eat them) can metabolize. But plants are, for the most part, rooted in one place.

Animals are mobile. They can move around and make use of any resources they can reach, whether they walk, run, jump, or fly. The plant must wait for rain. The animal can find water anywhere in its roaming range, or even migrate to find a new source. While the plant binds energy, the animal binds space.

Humans' social, imitative, and language abilities give us even more binding power. What makes humans special is that we can also bind *time*. We don't need to experience everything for ourselves over the course of a single lifetime. Instead, we benefit from the experiences of our predecessors, who can tell us what they've learned. Because we have evolved to imitate one another, a parent can show a child how to hunt, or how to operate the television remote. The child doesn't necessarily need to figure it out from scratch. Because we have evolved the ability to speak, we can use language to instruct others. Don't touch the red snake; it's poisonous.

Through language and instruction, humans create a knowledge base that compresses or binds many centuries of accumulated wisdom into the learning span of a single generation. We do not need to reinvent all knowledge anew, every time. But we must, at least provisionally, believe that people of the past have something to teach us.

14.

Imitation, social bonding, and language allowed humans to advance, each skill reinforcing the others. Happiness itself, research now suggests, is less the goal of social cohesiveness than an incentive—more like nature's bribe for us to play nicely with others. Even our emotions are not our own, but a side effect of how our social group is organized. The closer people are to the core of a social network, the happier they are. Happiness is not a function of one's individual experience or choice, but a property of groups of people.

Viewed this way, our emotions are simply triggers for new ties with others. One person is happy, and laughs. The laughter and emotion then spread from person to person throughout the network. The purpose may be less to spread happiness than to activate the network, reinforce connectivity, and coalesce the social group.

The reverse is also true. Disconnection from one's social group leads to higher rates of depression, illness, and mortality. A baby starved of social contact has difficulty developing a regulated nervous system. Young men with few social acquaintances develop high adrenaline levels. Lonely students have low levels of immune cells. Prison inmates prefer violence to solitary confinement. In the US, social isolation is a greater public health problem than obesity.

Being social may be the whole point. The things we learn from one another are helpful with the logistics of mutual survival, but the process of learning itself—the sense of connection, rapport, and camaraderie we develop while communicating— may be the greater prize. We may not socialize in order to live any more than we live in order to socialize.

15.

Of course, thriving, happy, connected people experience individuality also. We may be social, but we are also autonomous beings who enjoy exercising free will and independent choice.

Still, psychologists and social scientists recognize that the healthiest ways of expressing our autonomy occur within a larger social context. Making the independent choice to trust other people, or even to engage in self-sacrifice, allows people to feel that they are connected to a bigger project and acting

SOCIAL
ANIMALS

out of concern for the common good. Unfettered communications, a functioning democracy, the right to free expression and assembly, community values, and economic inclusion all enable such activities. Without a relatively open social landscape in which to participate, we can only express ourselves through self-absorption or withdrawal. We experience a limited sort of autonomy, like that of a child exercising independence by refusing to eat dinner.

This dynamic is self-reinforcing. Without socially positive opportunities to exercise our autonomy, we tend toward self-promotion over self-sacrifice and fixate on personal gain over collective prosperity. When we can't see ourselves as part of an enduring organism, we focus instead on our individual mortality. We engage in futile gestures of permanence, from acquiring wealth to controlling other people. We respond to collective challenges such as climate change through the self-preservation fantasy of a doomsday prepper. These limited attitudes trickle up through our political and consumer choices, making our social landscape less conducive to social cohesion.

Mental health has been defined as "the capacity both for autonomous expansion and for homonomous integration with others." That means our actions are governed from within, but directed toward harmonious interaction with the world. We may be driven internally, but all this activity is happening in relationship with our greater social environment. We can only express our autonomy in relationship to other people.

To have autonomy without interdependency leads to isolation or narcissism. To have interdependency with no autonomy stunts our psychological growth. Healthy people live in social groups that have learned to balance or, better, marry these two imperatives.

LEARNING
TO LIE

16.

It doesn't take much to tilt a healthy social landscape toward an individualist or repressive one. A scarcity of resources, a hostile neighboring tribe, a warlord looking for power, an elite seeking to maintain its authority, or a corporation pursuing a monopoly all foster antisocial environments and behaviors.

Socialization depends on both autonomy and interdependency, emphasizing one at the expense of the other compromises the balance.

For example, one desocializing strategy is to emphasize individualism. The social group is broken down into atomized individuals who fight for their right to fulfillment via professional advancement or personal consumption. This system is often sold to us as freedom. Those competing individuals never find true autonomy, however, because they lack the social fabric in which to exercise it.

Another path to desocialization emphasizes conformity. People don't need to compete because they are all the same. Such a system mitigates strident individualism, but it does so through obedience—often to a supreme ruler or monopoly party. Conformity is not truly social, however, because people are looking up for direction rather than to one another. There's no variation, mutation, or social fluidity, so conformity ends up being just as desocializing as individualism.

Both approaches depend on separating people from one another and undermining our evolved social mechanisms in order to control us.

17.

Any of our healthy social mechanisms can become vulnerabilities: what hackers would call "exploits" for those who want to manipulate us. When a charity encloses a free "gift" of return address labels along with their solicitation for a donation, they are consciously manipulating our ancient, embedded social bias for reciprocity. The example is trivial, but the pattern is universal. We either succumb to the pressure with the inner knowledge that something is off, or we recognize the ploy, reject the plea, and arm ourselves against such tactics in the future. In either case, the social landscape is eroded. What held us together now breaks us apart.

Indeed, the history of civilization can be understood by the ways in which we swing between social connection and utter alienation, and how our various media contribute to the process.

We invent new mechanisms for connection and exchange, from books and radio to money and social media. But then those very media become the means through which we are separated. Books reach only the literate wealthy; radio encourages mob violence; money is hoarded by monopoly bankers; social media divides users into algorithmically determined silos.

Unlike human beings ourselves, the media and technologies we develop to connect with one another are not intrinsically social.

18.

Spoken language could be considered the first communication technology. Unlike dilating pupils or mirror neurons, speech requires our conscious participation.

Language gave humans a big advantage over our peers, and

allowed us to form larger and better organized groups. Language bonded tribes, offered new ways to settle conflicts, allowed people to express emotions, and—maybe more important—enabled elders to pass on their knowledge. Civilization's social imperative could now advance faster than biology could engineer by itself.

But language also had the reverse effect. Before language, there was no such thing as a lie. The closest thing to lying would have been a behavior such as hiding a piece of fruit. Speech created a way of actively misrepresenting reality to others.

The written word, likewise, offered us the opportunity to begin recording history, preserving poetry, writing contracts, forging alliances, and sending messages to distant places. As a medium, it extended our communication across time and space, connecting people in ways previously unimaginable.

When we look at the earliest examples of the written word, however, we see it being used mostly to assert power and control. For the first five hundred years after its invention in Mesopotamia, writing was used exclusively to help kings and priests keep track of the grain and labor they controlled. Whenever writing appeared, it was accompanied by war and slavery. For all the benefits of the written word, it is also responsible for replacing an embodied, experiential culture with an abstract, administrative one.

The Gutenberg printing press extended the reach and accessibility of the written word throughout Europe, and promised a new era of literacy and expression. But the presses were tightly controlled by monarchs, who were well aware of what happens when people begin reading one another's books. Unauthorized presses were destroyed and their owners executed. Instead of promoting a new culture of ideas, the printing press reinforced control from the top.

Radio also began as a peer-to-peer medium. A radio set was

originally a transceiver—what we now think of as ham radio. As corporations lobbied to monopolize the spectrum and governments sought to control it, radio devolved from a community space to one dominated by advertising and propaganda.

Adolf Hitler used the new, seemingly magical medium of radio to make himself appear to be anywhere and everywhere at once. No single voice had so permeated German society before, and the sense of personal connection it engendered allowed Hitler to create a new sort of rapport with millions of people. The Chinese installed 70 million loudspeakers to broadcast what they called "Politics on Demand" throughout the nation. Rwandans used radio as late as 1993 to reveal the location of ethnic enemies so that mobs of loyalists with machetes could massacre them.

Once under the control of elites, almost any new medium starts to turn people's attention away from one another and toward higher authorities. This makes it easier for people to see other people as less than human and to commit previously unthinkable acts of violence.

Television was also originally envisioned as a great connector and educator. But marketing psychologists saw in it a way to mirror a consumer's mind and to insert within it new fantasies—and specific products. Television "programming" referred to the programmability not of the channel, but of the viewer. The bright box was captivating, perhaps even unintentionally capitalizing on embedded human instincts. Instead of sitting around the fire listening to one another's stories, we sat on the couch and stared into the screen. Group rapport was replaced with mass reception.

While television encouraged a conformist American culture through its depictions of family and a consumer utopia, it also pushed an equally alienating ethos of individualism. Television told people they could choose their own identities in the same way as they chose their favorite character on a soap opera. The viewing public gladly accepted the premise, and the social cost.

Television commercials depended on alienated individuals, not socially connected communities. A blue jeans commercial promising a sexier life doesn't work on someone already in a satisfying relationship. It is aimed at the person sitting alone on the couch. Television culture further fostered loneliness by substituting brand imagery for human contact.

Television was widely credited as the single biggest contributor to the desocialization of the American landscape, the decline of national clubs and community groups, and the sense of social isolation plaguing the suburban frontier.

That is, until the internet.

19.

The net seemed to offer personal autonomy at the same time as it connected people in new ways. Popular mythology holds that computer networks began as a sort of informational bomb shelter for the US Defense Department. In reality, computer networking started as a way of sharing processing power. It was a bit like cloud computing, where dumb terminals were connected to big but primitive mainframes. Processing cycles were scarce, and networking allowed many people to share the common resource.

An extra benefit of connected computers was the ability to leave messages. When your colleagues logged on, they'd see the little text files you left for them in their user folders. That was email. Messaging and other conferencing tools and bulletin boards soon became more popular than computing itself. Eventually, individual servers began connecting to one another, and what we think of today as networking was born.

The defense industry saw in these ad-hoc networks a new, resilient, and decentralized form of communication. If one part of a network was attacked, the rest could still function and even route around the broken parts. So the government funded the implementation of a big "network of networks" that finally became the internet.

But from the earliest days of networked computing, users were socializing, sharing recipes, or playing games instead of working. Although inhabited originally by scientists and defense contractors, the net soon became the province of cultural progressives, geeks, and intellectuals. The government didn't want it anymore and tried to sell it to AT&T, but even the communications company couldn't see the commercial possibilities of a free medium driven by the pleasure of communication.

For their part, the hackers and hippies inspired by the internet saw it as an extension of the human nervous system. Each human brain was understood to be a node in a giant network. Aspirations were high. The internet would transform humanity into the planet's brain; Gaia, the planet's spirit, was to become fully conscious.

Traditional media companies and advertisers, who had decidedly less interest in planetary consciousness than they did in quarterly profits, became gravely concerned when they learned in 1992 that the average internet-connected family was watching nine hours less commercial television per week than families without the internet. So they took a two-pronged approach, vilifying the net in their broadcasts and publications while also steering the internet toward less interactive and more advertiser-friendly uses.

The World Wide Web was originally intended as an easier way to find and hyperlink research documents. But its visual, clickable interface felt a lot more like television than the rest of the net, and attracted the interest of marketers. Users didn't need to type or actively think in order to participate. They could just click and read or, better, watch and buy.

To the dismay of the hippies and hackers building utopian virtual communities, the Web quickly became more of a shopping catalogue than a conversation space. Gone was connectivity between people, replaced by "one-to-one marketing" relationships between individuals and brands. Thousands of companies arose to peddle their wares in the dotcom boom—more companies than could possibly make a profit, leading to the dotcom bust.

Internet utopians declared victory: the net had survived an attack from the forces of commercialization and could now resume its mission to connect us all. We announced that the net was and would always be a "social medium."

20.

Social media began with the best of intentions. In the wake of the dotcom bust, after investors had declared the internet "over," a new generation of developers began building publishing tools that defied the Web's top-down, TV-like presentation and instead let the people make the content. Or *be* the content.

At the time, it felt revolutionary. With social media, the net could eschew commercial content and return to its amateur roots, serving more as a way for people to find one another, forge new alliances, and share unconventional ideas.

The new blogging platforms that emerged let users create the equivalent of web pages, news feeds, and discussion threads, instantly. A free account on a social media platform became an instant home base for anyone. The simple templates allowed people to create profiles, link to their favorite music and movies, form lists of friends, and establish a new, if limited, outlet for self-expression in a global medium.

The emphasis of these platforms changed, however, as the companies behind them sought to earn money. Users had grown accustomed to accessing internet services for free in return for looking at some advertisements. In the social media space, however, ads could be tightly targeted. The copious information that users were posting about themselves became the basis for detailed consumer profiles that, in turn, determined what ads reached which users.

The experience of community engendered by social media was quickly overtaken by a new bias toward isolation. Advertisers communicated individually to users through news feeds that were automatically and later algorithmically personalized. Even this wasn't understood to be so bad, at first. After all, if advertisers are subsidizing the community platform, don't they deserve a bit of our attention? Or even a bit of our personal information? Especially if they're going to work hard to make sure their ads are of interest to us?

What people couldn't or wouldn't pay for with money, we would now pay for with personal data. But something larger had also changed. The platforms themselves were no longer in the business of delivering people to one another; they were in the business of delivering people to marketers. Humans were no longer the customers of social media. We were the product.

In a final co-option of the social media movement, platforms moved to turn users into advertisers. Instead of bombarding people with messages from companies, social media platforms pushed an online upgrade on word-of-mouth they called "social recommendations." Some marketers worked to get people to share links to ads and content they liked. Others sought out particularly influential users and made them brand advocates who got paid in free product.

From then on, the members of various affinity groups and even political affiliations competed against one another for likes, followers, and influencer status. These metrics came to matter not just as an interesting measure of one's social influence but as a way of qualifying for sponsorships, roles in music videos, speaking invitations, and jobs.

The social agenda driving what felt like a new medium's natural evolution was once again subsumed by competitive individualism.

21.

An increasingly competitive media landscape favors increasingly competitive content. Today, anyone with a smartphone, web page or social media account can share their ideas. If that idea is compelling, it might be replicated and spread to millions. And so the race is on. Gone are the collaborative urges that characterize embodied social interaction. In their place comes another bastardized Darwinian ideal: a battle for survival of the fittest meme.

The term "media virus" was meant to convey the way ideas could spread in a world with more interactive communications. A real, biological virus has a novel, never-before-seen protein shell that lets it travel through a person's bloodstream unrecognized. (If the body identifies the virus, it sends antibodies to attack it.) The virus then latches onto a cell in the host organism and injects its genetic code within. The code works its way to the nucleus of the cell and seeks to interpolate itself into the cell's DNA. The next time the cell reproduces, it replicates the virus's code along with its own.

Then the person carrying the virus begins spreading it to others. The virus continues to replicate and spread until, at last, the body learns to reject its code. From then on, the protein shell will be recognized and attacked, even if it comes back months or years later. Immunity.

A media virus works the same way. It has a novel, unrecognizable shell—a unique, rule-breaking use of media that seems so sensational we can't help but spread it. A woman "live streams" her husband dying of gunshot wounds. A congressman transmits smartphone pictures of his genitals to a minor. A president threatens a nuclear attack in a public, 140-character message typed with his thumbs.

In each case, the story's initial proliferation has more to do with the medium than the message. The viral shell is not just a media phenomenon, but a way of grabbing attention and paralyzing a person's critical faculties. That moment of confusion creates the time and space for infection.

The virus continues to replicate only if its code can successfully challenge our own. That's why the ideas inside the virus—the memes—really matter. A fatal car crash attracts our attention because of the spectacle, but worms its way into our psyche because of our own conflicted relationship with operating such dangerous machinery, or because of the way it disrupts our ongoing denial of our own mortality.

Likewise, a contagious media virus attracts mass attention for its spectacular upending of TV or the net, but then penetrates the cultural psyche by challenging collectively repressed anxieties. Surveillance video of a police van running over a black suspect recalls America's shamefully unacknowledged history of slavery and ongoing racism. The social media feed of a neo-Nazi bot in Norway can stimulate simmering resentment of the European Union's dissolution of national identities.

DOUGLAS RUSHKOFF

The amazing thing is that it doesn't matter what side of an issue people are on for them to be infected by the meme and provoked to replicate it. "Look what this person said!" is reason enough to spread it. In the contentious social media surrounding elections, the most racist and sexist memes are reposted less by their advocates than by their outraged opponents. That's because memes do not compete for dominance by appealing to our intellect, our compassion, or anything to do with our humanity. They compete to trigger our most automatic impulses.

22.

We can't engineer a society through memetics the way a biologist might hope to engineer an organism through genetics. To do so would bypass our higher faculties, our reasoning, and our collective autonomy. It is unethical, and, in the long run, ineffective. It's also deliberately antihuman.

Sure, well-meaning and pro-social counterculture groups have attempted to spread their messages through the equivalents of viral media. They subvert the original meanings of corporate logos, leveraging the tremendous spending power of an institution against itself with a single clever twist. With the advent of a new, highly interactive media landscape, internet viruses seemed like a great way to get people talking about unresolved issues. If the meme provokes a response, this logic argues, then it's something that *has* to be brought to the surface.

The problem is, the ends don't always justify the memes. Today, the bottom-up techniques of guerrilla media activists are in the hands of the world's wealthiest corporations, politicians, and propagandists. To them, viral media is no longer about revealing inequality or environmental threats. It's simply an effective means of generating a response, even if that response is automatic, unthinking, and brutish. Logic and truth have nothing to do with it. Memes work by provoking fight-or-flight reactions. Those sorts of responses are highly individualist. They're not prosocial, they're antisocial.

Not that the technique was ever appropriate, even practiced benevolently. The danger with viruses is that they are constructed to bypass the neocortex—the thinking, feeling part of our brain—and go straight to the more primal brain stem beneath. The meme for scientifically proven climate change, for example, doesn't provoke the same intensity of cultural response as the meme for "elite conspiracy!" A viral assault will not persuade a flood-ravaged town to adopt strategies of mutual aid. It could, on the other hand, help push survivors toward more paranoid styles of self-preservation. Memetic campaigns do not speak to the part of the brain that understands the benefits of tolerance, social connection, or appreciation of difference. They're speaking to the reptile brain that understands only predator or prey, fight or flight, kill or be killed.

23.

Memetics, the study of how memes spread and replicate, was first popularized by an evolutionary biologist in the 1970s. A strident atheist, the scientist meant to show how human culture evolves by the same set of rules as any other biological system: competition, mutation, and more competition. Nothing special going on here.

It turns out there *is* something special going on here, and that there are a few things missing from this simplistic explanation of memes and genes alike. Neither genes nor memes determine *everything* about an organism or a culture. DNA is not a static blueprint but acts differently in different situations. Genes matter, but the *expression* of those genes matters more. Expression is entirely dependent on the environment, or the "protein soup" in which those genes are swimming. It's why a tame grasshopper can, under the right conditions, transform into a rapacious, swarming locust.

Genes are not solo actors. They do not selfishly seek their own replication at all costs. Newer science describes them as almost social in nature: organisms get information from the environment and one another for how to change. The conditions, the culture, and connectivity are as important as the initial code.

Likewise, memes are not interacting in an ideological vacuum. If we truly want to understand cultural contagion, we must place equal importance on the memes, the viral shell around those memes, and the ideological soup in which those memes attempt to replicate. Early memeticists saw memes as competing against one another, but that's not quite right. Memes trigger and exploit our untapped fear, anxiety, and rage in order to manipulate us. They are not attacking one another; they are attacking us humans.

It's not the meme that matters, but the culture's ability to muster an effective immune response against it.

24.

The technologies through which memes are being transmitted change so rapidly that it is impossible to recognize their new forms—their shells—in advance. We must instead build our collective immune system by strengthening our organic coherence—our resistance to the socially destructive memes within them.

This is particularly difficult when the enemies of Team Human are busy escalating memetic warfare with artificial intelligence. Each of their algorithms is designed to engage with us individually, disconnect us from one another, neutralize our defense mechanisms, and program our behavior as if we were computers. Television advertisers may have normalized the idea that consumers can be experimented on like lab rats, but social media weaponizes these techniques.

At least television happened in public. The enormity of the audience was the source of its power, but also its regulating force. TV stations censored ads they found offensive for fear of losing viewers. In contrast, social media messages may cost pennies or nothing at all, are seen only by the individuals who have been targeted, and are placed by bots with no qualms about their origins or content.

Moreover, when media is programmed to atomize us and the messaging is engineered to provoke our most competitive, reptilian sensibilities, it's much harder to muster a collective defense. We lose our ability to distinguish the real from the unreal, the actual from the imagined, or the threat from the conspiracy.

The powers working to disrupt democratic process through memetic warfare understand this well. Contrary to popular accounts, they invest in propaganda from all sides of the political spectrum. The particular memes they propagate through social media are less important than the immune reactions they hope to provoke.

Memetic warfare, regardless of the content, discourages cooperation, consensus, or empathy. The reptile brain it triggers doesn't engage in those prosocial behaviors. Instead, in an environment of hostile memes and isolated by social media, human beings become more entrenched in their positions and driven by a fear for their personal survival. Worst of all, since these platforms appear so interactive and democratic, we experience this degradation of our social processes as a form of personal empowerment. To be truly social starts to feel like a restraint—like the yoke of political correctness, or a compromising tolerance of those whose very existence weakens our stock.

This may not have been the intent of social media or any of the communications technologies that came before it—all the way back to language. The internet doesn't have to be used against a person's critical faculties any more than we have to use language to lie or written symbols to inventory slaves. But each extension of our social reality into a new medium requires that we make a conscious effort to bring our humanity along with us.

We must protect our social human organism from the very things we have created.

FIGURE AND GROUND

25.

Human inventions often end up at cross purposes with their original intentions—or even at cross purposes with humans, ourselves. Once an idea or an institution gains enough influence, it changes the basic landscape. Instead of the invention serving people in some way, people spend their time and resources serving *it*. The original subject becomes the new object.

Or, as we may more effectively put it, the figure becomes the ground.

The idea of figure and ground was first posited by a Danish psychologist in the early 1900s. He used a simple cardboard cutout to test whether people see the central image or whatever is around it. We now know the experiment as the drawing that can be seen as a white vase if you look at the middle of the image, or as two black faces in profile if you focus on the periphery. The model of perception was useful to psychologists, who were attempting to understand how the brain identifies and remembers things.

What fascinates people to this day is the way the perception of figure or ground can change in different circumstances and cultures. When shown a picture of a cow in a pasture, most westerners will see a picture of a cow. Most easterners, on the other hand, will see a picture of a pasture. Their perceptions are so determined, in fact, that people who see the figure may be oblivious to major changes in the background, and people who see the ground may not even remember what kind of animal was grazing there.

Neither perception is better or worse, so much as incomplete. If the athlete sees herself as the only one that matters, she misses the value of her team—the ground in which she functions. If a company's "human resources" officer sees the individual employee as nothing more than a gear in the firm, he misses the value and autonomy of the particular person, the figure.

When we lose track of figure and ground, we forget who is doing what for whom, and why. We risk treating other people as objects. Worse, we embed these values in our organizations or encode them into our technologies. By learning to recognize reversals of figure and ground, we can liberate ourselves from the systems to which we have become enslaved.

26.

Figure/ground reversals are easy to spot once you know where to look, and maybe *how* to look.

Take money: it was originally invented to store value and enable transactions. Money was the medium for the marketplace's primary function of value exchange. Money was the ground, and the marketplace was the figure. Today, the dynamic is reversed: the acquisition of money itself has become the central goal, and the marketplace just a means of realizing that goal. Money has become the figure, and the marketplace full of people has become the ground.

Understanding this reversal makes it easier to perceive the absurdity of today's destructive form of corporate capitalism. Corporations destroy the markets on which they depend, or sell off their most productive divisions in order to increase the bottom line on their quarterly reports. That's because the main product of a company is no longer whatever it provides to consumers, but the shares it sells to investors. The figure has become the ground.

Or consider the way the human ideal of education was supplanted by its utilitarian opposite. Public schools were originally conceived as a way of improving the quality of life for workers. Teaching people to read and write had nothing to do with making them better coal miners or farmers; the goal was to expose the less privileged classes to the great works of art, literature, and religion. A good education was also a requirement for a functioning democracy. If the people don't have the capacity to make informed choices, then democracy might easily descend into tyranny.

Over time, as tax dollars grew scarce and competition between nations fierce, schools became obliged to prove their value more concretely. The Soviets' launch of the Sputnik satellite in the 1960s led America to begin offering advanced math in high school. Likewise, for the poor in particular, school became the ticket to class mobility. Completing a high school or college education opens employment opportunities that would be closed otherwise—another good, if utilitarian, reason to get educated.

But once we see competitive advantage and employment opportunity as the primary purposes of education rather than its ancillary benefits, something strange begins to happen. Entire curriculums are rewritten to teach the skills that students will need in the workplace. Schools consult corporations to find out what will make students more valuable to them. For their part, the corporations get to externalize the costs of employee training to the public school system, while the schools, in turn, surrender their mission of expanding the horizons of the working class to the more immediate purpose of job readiness.

Instead of compensating for the utilitarian quality of workers' lives, education becomes an extension of it. Where learning was the purpose—the figure—in the original model of public education, now it is the ground, or merely the means through which workers are prepared for their jobs.

27.

Technologies seem free of embedded agendas. Their automation and opacity make highly reversed situations appear quite normal and natural.

For example, most Americans accept the premise that they need a car to get to work. And a better car leads to a more pleasant commute. But that's only because we forgot that our pedestrian and streetcar commutes were forcibly dismantled by the automobile industry. The geography of the suburban landscape was determined less by concern for our quality of life than to promote the sales of automobiles that workers would be required to use. The automobile made home and work less accessible to each other, not more—even though the car looks like it's enhancing the commute.

Once the figure and ground have been reversed, technology only disguises the problem.

In education, this takes the form of online courses that promise all the practical results for a fraction of the cost and inconvenience. The loftier goals of learning or enrichment are derided as inefficient self-indulgence, or the province of decadent elites. Online courses don't require a campus or even a live teacher. A responsive, algorithmically generated curriculum of videos and interactive lessons is customized for the individual learner. It is the pinnacle of utilitarian education. Particular learning outcomes are optimized, and skills acquisition can be assessed through testing—also via computer.

Of course, when an online company is assessing its own efficacy, it is bound to make positive determinations. Even when companies don't tip the scales in their own favor, the underlying technology will be biased toward measuring only the educational values that have been programmed into it.

Automated classes work for rudimentary, job-related skills such as machine repair, simple medical procedures, or data entry. But they are terrible for creative thinking and interpretation. They don't even work particularly well for learning computer code, which is why most serious developers end up abandoning online code schools in favor of real-life programming boot camps with human instructors and struggling peers. People who learn by computer are not encouraged to innovate. They are simply being trained to do a job. They can repeat tasks, but they can't analyze or question the systems to which they are contributing. The more that real colleges incorporate the methods of their online competition, the less of this deeper learning they can offer.

For these reasons, many of the most ambitious engineers, developers, and entrepreneurs end up dropping out of college altogether. One tech billionaire famously offers $100,000 to twenty young people each year to abandon college in order to pursue their own ideas. The message he's sending to students is clear: if you want to get somewhere significant, don't worry about school.

When we reduce education to its utilitarian function, it may as well be performed by computers. Besides, as the anti-education billionaire would argue, people learn job skills better on the job itself—as interns or entry-level employees. But people who so easily dismiss education have forgotten what school is really *for*.

A live educator offers more than the content of a course. Human interaction and presence are important components of effective pedagogy. Moreover, a teacher sets an example by embodying the ideals of learning and critical thinking. Possessed by a spirit of inquiry, the teacher enacts the process of learning for students to mimic. The act of mimesis itself matters: one human learning by watching another, observing the subtle details, establishing rapport, and connecting to history. It's the ancient practice of people imitating people, finding role models, and continuing a project from one generation to the next.

The human social engagement is the thing; the utilitarian applications are just the excuse. When we allow those two to be reversed, the figure becomes the ground.

28.

Under the pretense of solving problems and making people's lives easier, most of our technological innovations just get people out of sight, or out of the way. This is the true legacy of the Industrial Age.

Consider Thomas Jefferson's famous invention, the dumbwaiter. We think of it as a convenience: instead of carrying food and wine from the kitchen up to the dining room, the servants could place items into the small lift and convey it upstairs by pulling on ropes. Food and drink appeared as if by magic. But the purpose of the dumbwaiter had nothing to do with saving effort. Its true purpose was to hide the grotesque crime of slavery.

This may be less technology's fault than the way we've chosen to use it. The Industrial Age brought us many mechanical innovations, but in very few cases did they actually make production more efficient. They simply made human skill less important, so that laborers could be paid less. Assembly line workers had to be taught only a single, simple task, such as nailing one tack into the sole of a shoe. Training took minutes instead of years, and if workers complained about their wages or conditions, they could be replaced the next day.

The industrialist's dream was to replace them entirely—with machines. The consumers of early factory goods loved the idea that no human hands were involved in their creation. They marveled at the seamless machined edges and perfectly spaced stitches of Industrial Age products. There was no trace of humans at all.

Even today, Chinese laborers "finish" smartphones by wiping off any fingerprints with a highly toxic solvent proven to shorten the workers' lives. That's how valuable it is for consumers to believe their devices have been assembled by magic rather than by the fingers of underpaid and poisoned children. Creating the illusion of no human involvement actually costs human lives.

Of course, the mass production of goods requires mass marketing—which has proven just as dehumanizing. While people once bought products from the people who made them, mass production separates the consumer from the producer, and replaces this human relationship with the brand. So where people used to purchase oats from the miller down the block, now consumers go to the store and buy a box shipped from a thousand miles away. The brand image—in this case, a smiling Quaker—substitutes for the real human relationship, and is carefully designed to appeal to us more than a living person could.

To pull that off, producers turned again to technology. Mass production led to mass marketing, but mass marketing required mass media to reach the actual masses. We may like to think that radio and TV were invented so that entertainers could reach bigger audiences, but the proliferation of broadcast media was subsidized by America's new, national brands, which gained access to consumers coast to coast. Marketers of the period believed they were performing a patriotic duty: fostering allegiance to mass-market brands that iconized American values and ingenuity. But the collateral damage was immense.

Consumer culture was born, and media technologies became the main way to persuade people to desire possessions over relationships and social status over social connections. The less fruitful the relationships in a person's life, the better target that person was for synthetic ones. The social fabric was undone.

Since at least the Industrial Age, technology has been used as a way to make humans less valued and essential to labor, business, and culture. This is the legacy that digital technology inherited.

29.

In a digital media environment, it's particularly easy for figure and ground—people and their inventions—to flip roles. Because so much of this landscape is programmed to do one thing or another, human beings often become the more passive, automatic players compared with the code that actively defines the terrain and influences our behaviors.

If we don't truly know what something is programmed to do, chances are it is programming *us*. Once that happens, we may as well be machines ourselves.

For example, memetic warfare treats the figure—the human—as ground. In a digital media environment, memes become more than catchy slogans or ideas; they are a form of code, engineered to infect a human mind and then to turn that person into a replicator of the virus. The meme is software, and the person is the machine.

The meme has one goal: to be reproduced. It enters the human, inciting confusion, excitement, panic, or rage and stimulating the host to pass it on. The meme is issuing one command: make me. And whether by word of mouth, social media, or viral video link, whether in agreement with the meme or in outraged rejection, the infected human obeys.

This automation tends to reinforce itself. The more we experience people as dehumanized replicators of memes, the more likely we are to treat one another as machines to be operated rather than as peers with whom to collaborate. Combined with other styles of manipulation and triggering, we end up in a world where success in business, politics, or even dating appears to depend on our ability to control others. Our essentially social activities become entirely instrumentalized, losing connection to their greater purpose. Under the pretense of optimizing probabilities or gaining broader choice, we surrender real connection and meaning. We become mere instruments.

The big reversal here is that our technologies used to be the instruments. They were extensions of our will—expressions of our autonomy as human beings. They gave us more choices. Of course, each new choice our technologies opened up to us also risked alienating us from our primal connections with nature and one another. Fire let us live in places otherwise too cold for human habitation. Electric lights let us stay up and do things late into the night. Airplanes let us travel across a dozen time zones in a single day. Sedatives let us sleep on the plane, stimulants wake us up when we arrive, and mood drugs help us cope with the stress of living that way. Sunrise and sunset are images for the computer desktop.

As we drift away from the biological clocks through which we used to find coherence, we become more dependent on artificial cues. We begin living as if we were in a shopping mall or casino, where day and night—as well as desire—are programmed by the environment. Everything is *strategized* by something or someone, even though the walls, lights, ceiling, and signage appear like features of the natural world. We are being optimized by something outside ourselves, toward purposes we don't even know. The Muzak in the supermarket is programmed to increase the rhythm at which we place things into our shopping carts, while the lighting in the office changes to increase our productivity during the afternoon "lull."

Our digital world is like the ultimate casino in this respect. It may have begun as a series of tools for specific purposes—spreadsheets, word processors, calculators, messaging, calendars, contact lists—but these tools went from being analogs or metaphors for real-life activities into being their replacements. Our technologies change from being the tools humans use into the environments in which humans function.

Think of the way video game graphics advanced from crude vectors indicating spacecrafts or asteroids to high-resolution, texture-mapped simulations of worlds. Playability has never depended on realism, any more than an authentic reproduction of a gun makes for a better game of cops and robbers than a stick. The more realistically a play world is depicted, the less play is involved and the more easily the player is manipulated to spend more time, energy, or money in the alternate world. As gaming goes from toy to simulation, the player becomes the played. Similarly, as technology goes from tool to replacement, the humans using it devolve from users to the used.

Yes, the digital world offers more choices—but it's not the humans, or at least not the users, who get to make them.

30.

Those of us who witnessed the dawn of the interactive era originally believed that digital technology was going to bring back the human powers and priorities that industrialization had vanquished.

The earliest interactive tools, such as the television remote control, changed our relationship to programming. Where breaking the captive spell of television used to require walking up to the set and physically turning a dial, the remote let us escape with the micro-motion of a single finger. As cable television expanded the offerings, we found ourselves less watching a particular TV program than playing the TV itself: moving from channel to channel and keeping track of multiple shows, or observing the similarities and contrasts between them.

Taking this play, the joystick turned the television into a game console. As thrilling as Pong or Space Invaders may have been, the mere ability to move the pixels on the screen felt revolutionary. Just as the remote control let us deconstruct the content, the joystick demystified the technology. The pixel was no longer the exclusive province of corporate news networks and Hollywood stars, but something that could be manipulated by anyone. Likewise, the video recorder turned viewers into producers, breaking the monopoly on screen space.

Finally, the computer keyboard and mouse turned the TV from a monitor into a portal. The internet gave us all the ability to share, distribute our own media, and promote ideas from the bottom up. At last, a medium had arisen that seemed to connect people rather than alienate us from one another. The content became less important than the contact. The internet would serve as remedial help for a society desocialized by commercial television.

Unfortunately, the people and companies that were still heavily invested in Industrial Age values sought to undo the liberating impact of the remote control, the joystick, and the mouse. The technology industry wasn't consciously attacking human autonomy so much as reinforcing its users' role as passive consumers from whom to extract value.

Internet industry magazines declared that we were living in an "attention economy," where a company's profits would depend on its ability to wrest "eyeball hours" from users. Kids using the remote to escape commercials and surf manipulative content from a safe distance were accused of having dangerously shortened attention spans. But how else were children supposed to respond to a world in which every place they set their gaze was an advertisement? As the number of amphetamine prescriptions for young people continues to double every few years, we must at least consider the environmental factors that have contributed to widespread attention deficit, and whether we have been indiscriminately drugging some young people into compliance.

Meanwhile, computer interfaces became unnecessarily inaccessible. Early computers could be controlled with simple typed commands. Learning to use a computer was the same thing as learning to program a computer. Yes, it took a couple of hours, but it put the user in a position of authority over the machine. If a program couldn't do something, a user knew whether this was because the computer was incapable of that function or because the programmer simply didn't want the user to be able to do it.

In an effort to make computers easier to *use*, developers came up with elaborate metaphors for desktops and files. Consumers gained a faster onramp to using a computer, but were further distanced from programming. One operating system required users to summon "the Wizard" in order to install software; it was surely meant as a friendly help feature, but the choice of a wizard underscored just how mysterious the inner workings of an application directory were supposed to be to a typical user.

Finally, the new culture of contact enabled by digital networks was proving unprofitable and was replaced by an industry-wide ethos of "content is king." Of course, content was not the message of the net; the social contact was. We were witnessing the first synaptic transmissions of a collective organism attempting to reach new levels of connectedness and wake itself up. But that higher goal was entirely unprofitable, so conversations between actual humans were relegated to the comments sections of articles or, better, the reviews of products. If people were going to use the networks to communicate, it had better be about a brand. Online communities became affinity groups, organized around purchases rather than any sort of mutual aid. Actual "social" media was only allowed to flourish once the contacts people made with one another became more valuable as data than the cost in missed shopping or viewing time.

Content remained king, even if human beings were now that content.

THE
DIGITAL
MEDIA
ENVIRONMENT

31.

We were naive to think that digital technology would be intrinsically and inevitably more empowering than any medium that came before it. Yes, digital networks are more directionless and decentralized than their broadcast predecessors. They allow messages to flow from the bottom up, or the outside in. But, like all media, if they're not consciously seized by the people seeking empowerment, they'll be seized by someone or something else.

Whoever controls media controls society.

Each new media revolution appears to offer people a new opportunity to wrest that control from an elite few and reestablish the social bonds that media has compromised. But, so far anyway, the people—the masses—have always remained one entire media revolution behind those who would dominate them.

For instance, ancient Egypt was organized under the presumption that the pharaoh could directly hear the words of the gods, as if he were a god himself. The masses, on the other hand, could not hear the gods at all; they could only believe.

With the invention of text, we might have gotten a literate culture. But text was used merely to keep track of possessions and slaves. When writing was finally put in service of religion, only the priests could read the texts and understand the Hebrew or Greek in which they were composed. The masses could hear the Scriptures being read aloud, thus gaining the capability of the prior era—to hear the words of God. But the priests won the elite capability of literacy.

When the printing press emerged in the Renaissance, the people gained the ability to read, but only the king and his cho-

sen allies had the power to produce texts. Likewise, radio and television were controlled by corporations or repressive states. People could only listen or watch.

With computers came the potential to program. Thanks to online networks, the masses gained the ability to write and publish their own blogs and videos—but this capability, writing, was the one enjoyed by the elites in the prior revolution. Now the elites had moved up another level, and were controlling the software through which all this happened.

Today, people are finally being encouraged to learn code, but programming is no longer the skill required to rule the media landscape. Developers can create the apps they want, but its operation and distribution are entirely dependent on access to the walled gardens, cloud servers, and closed devices under the absolute control of just three or four corporations. The apps themselves are merely camouflage for the real activity occurring on these networks: the hoarding of data about all of us by the companies that own the platforms.

Just as with writing and printing, we believe we have been liberated by the new medium into a boundless frontier, even though our newfound abilities are entirely circumscribed by the same old controlling powers. At best, we are settling the wilderness for those who will later monopolize our new world.

32.

The problem with media revolutions is that we too easily lose sight of what it is that's truly revolutionary. By focusing on the shiny new toys and ignoring the human empowerment potentiated by these new media—the political and social capabilities they are retrieving—we end up surrendering them to the powers

that be. Then we and our new inventions become mere instruments for some other agenda.

Social phenomena of all sorts undergo this process of hollowing. When punk rockers reduce their understanding of their movement to the right to wear Mohawks or pierce their faces, it's easy for them to lose touch with the more significant anti-authoritarian ideology of DIY, direct action, and never selling out. Instead, punk becomes just another fashion trend to be sold at the mall. When ravers understand their movement as the right to take drugs and dance all night, they lose sight of the deeper political potentials unleashed by reclaiming public space or separating recreation from profit. Rave becomes just another genre for industry to sell. The styles of these movements were co-opted, and the essential shifts in power on which they were based were left behind.

With digital technology, we too quickly let go of the social and intellectual empowerment offered by these new tools, leaving them to become additional profit centers for the already powerful. For example, the early internet enabled new conversations between people who might never have connected in real life. The networks compressed distance between physicists in California, hackers in Holland, philosophers in Eastern Europe, animators in Japan—and this writer in New York.

These early discussion platforms also leveraged the fact that, unlike the TV or telephone, internet messaging didn't happen in real time. Users would download net discussions, read them in their own time, offline, and compose a response after an evening of thought and editing. Then they would log back onto the net, upload the contribution, and wait to see what others thought.

As a result, the internet became a place where people sounded and acted smarter than they do in real life. Imagine

that: a virtual space where people brought their best selves, and where the high quality of the conversation was so valued that communities governed these spaces the way a farmers' cooperative protects a common water supply. To gain access to the early internet, users had to digitally sign an agreement not to engage in any commercial activity. Advertising was expressly forbidden. Even the corporate search and social platforms that later came to monopolize the net originally vowed never to allow advertising because it would taint the humanistic cultures they were creating.

Over time, enthusiasm for the intellectual purity of the net was overtaken by the need to appeal to investors. Business magazines announced that the internet could save the dying stock market by creating more room for the economy to grow—even if that additional real estate was virtual. A search engine designed to promote academic thought became the world's biggest advertising agency, and a social media platform designed to help people connect became the world's biggest data collector.

Enthusiasts still associated the net with education and political power. They pushed for technology in schools and laptops in Africa, even though the digital society's essential values had been left behind in the era of 2400-baud modems. The primary purpose of the internet had changed from supporting a knowledge economy to growing an attention economy. Instead of helping us leverage time to our intellectual advantage, the internet was converted to an "always on" medium, configured to the advantage of those who wanted to market to us or track our activities.

Going online went from an active choice to a constant state of being. The net was strapped to our bodies in the form of smartphones and wearables that can ping or vibrate us to attention with notifications and updates, headlines and sports scores,

social media messages and random comments. We have ended up living in a state of perpetual interruption that used to be endured only by 911 emergency operators or air traffic controllers, only we do it 24/7 and we pay for the privilege.

The resulting disorientation is self-reinforcing. The more we are interrupted, the more distracted we become, and the less we avail ourselves of the real-world markers we use to ground ourselves. We become more easily manipulated and directed by the many technologies whose very purpose is to disorient us and control our behavior.

We humans go from being the figure in a digital environment to being the ground.

33.

Living in a digitally enforced attention economy means being subjected to a constant assault of automated manipulation. Persuasive technology, as it's now called, is a design philosophy taught and developed at some of America's leading universities and then implemented on platforms from e-commerce sites and social networks to smartphones and fitness wristbands. The goal is to generate "behavioral change" and "habit formation," most often without the user's knowledge or consent.

Behavioral design theory holds that people don't change their behaviors because of shifts in their attitudes and opinions. On the contrary, people change their attitudes to match their behaviors. In this model, we are more like machines than thinking, autonomous beings. Or at least we can be made to work that way.

That's why persuasive technologies are not designed to influence us through logic or even emotional appeals. This isn't advertising or sales, in the traditional sense, but more like war-

time psy-ops, or the sort of psychological manipulation exercised in prisons, casinos, and shopping malls. Just as the architects of those environments use particular colors, soundtracks, or lighting cycles to stimulate desired behavior, the designers of web platforms and phone apps use carefully tested animations and sounds to provoke optimal emotional responses from users. Every component of a digital environment is tested for its ability to generate a particular reaction, be it more views, more purchases, or just more addiction. New mail is a happy sound; no mail is a sad one. The physical gesture of swiping to update a social media feed anchors and reinforces the compulsive urge to check in, just in case.

Most persuasive technology tactics depend on users trusting that a platform accurately represents the world it claims to depict. The more we accept the screen as a window on reality, the more likely we are to accept the choices it offers. But what of the choices that it doesn't offer? Do they really not exist?

A simple search for "a pizzeria near me" may list all the restaurants that have paid to be found, but not those that haven't. Persuasive designs offer users options at every juncture, in order to simulate the experience of choice without the risk of the user exercising true autonomy and wandering off the reservation. It's the same way game designers lead all players through the same game story, even though we feel like we're making independent choices from beginning to end. None of these choices are real, because every one leads inevitably to the outcome that the designers have predetermined for us. Because the interfaces look neutral, we accept the options they offer at face value. The choices are not choices at all, but a new way of getting us to accept limitations. Whoever controls the menu controls the choices.

Designers seeking to trigger addictive loops utilize what

behavioral economists call "variable rewards." Marketing psychologists in the 1950s discovered that human beings are desperate to figure out the pattern behind getting rewarded. It makes sense from a survival perspective: sudden wind gusts mean rain is coming, and predicting the motions of a fish in the water makes it easier to spear.

In the hands of slot machine designers, this survival trait becomes an exploit. By doling out rewards at random intervals, the slot machine confounds the human being attempting to figure out the pattern. Consciously, we know the machine is randomized or perhaps even programmed unfairly to take our money. But subconsciously, those random intervals of quarters clinking into the metal tray induce a compulsive need to keep going. Are the quarters coming down every ten tries? Every five tries? Or is it five, seven, ten, and then back to five? Let me try again and see . . .

Compulsions are futile efforts at gaining control of random systems. Once triggered, they are really hard to shake. That's why academic studies of slot machine patterning became required reading in the euphemistically named "user experience" departments of technology companies. It's how we've become addicted to checking our email—looking to see if there's just one rewarding email for every ten useless ones. Or is it every eleven?

Persuasive design also exploits our social conditioning. We evolved the need to be aware of anything important going on in our social circle. Not having the knowledge that a fellow group member is sick or angry could prove disastrous. In the compliance professional's hands, this "fear of missing out" provides direct access to our behavioral triggers. All we need are a few indications that people are talking about something to stimulate our curiosity and send us online and away from whatever we were really doing. So designers put a red dot with a number in it over an app's icon to make sure we know that something's happening, comments are accumulating, or a topic is trending. If you refuse to heed the call, you may be the last person to find out.

User experiences are also designed to trigger our social need to gain approval and meet obligations—ancient adaptations for group cohesion now turned against us. We seek high numbers of "likes" and "follows" on social platforms because these metrics are the only way we have of gauging our social acceptance. There's no difference in intensity between them. We can't know if we are truly loved by a few; we can only know if we got liked by many. Similarly, when someone likes something of ours or asks to be our "friend" on a social network, we feel socially obligated to return the favor.

Some platforms leverage our urge to compete, even playfully, with our peers. Websites employ leader boards to honor those who have made the most posts, trades, or whatever metric a company wants to promote. Users also compete for "streaks" of daily participation, badges of accomplishment, and other ways of demonstrating their achievements and status—even if just to themselves. Gamification is used to motivate employees, students, consumers, and even online stock traders. But in the process, the play is often surrendered to some pretty unplayful outcomes, and people's judgment is clouded by the unrelated urge to win.

Many familiar interfaces go even deeper, building addictive behaviors by leveraging our feeding instincts. Designers have discovered that "bottomless feeds" tend to keep users swiping down for additional articles, posts, or messages, consuming more than they intend to because of that unsatiated feeling of never reaching the end.

On the other hand, designers want to keep us in a state

of constant disorientation—always scrolling and paying attention to something, but never so much attention that we become engrossed and regain our bearings. So they use interruption to keep us constantly moving from one feed to another, checking email and then social media, videos, the news and then a dating app. Each moment of transition is another opportunity to offer up another advertisement, steer the user toward something yet more manipulative, or extract more data that can be used to profile and persuade more completely.

34.

Instead of designing technologies that promote autonomy and help us make informed decisions, the persuasion engineers in charge of our biggest digital companies are hard at work creating interfaces that thwart our cognition and push us into an impulsive state where thoughtful choices—or thought itself—are nearly impossible.

We now know, beyond any doubt, that we are dumber when we are using smartphones and social media. We understand and retain less information, comprehend with less depth, and make decisions more impulsively than we do otherwise. This untethered mental state, in turn, makes us less capable of distinguishing the real from the fake, the compassionate from the cruel, and even the human from the nonhuman. Team Human's real enemies, if we can call them that, are not just the people who are trying to program us into submission, but the algorithms they've unleashed to help them do it.

Algorithms don't engage with us humans directly. They engage with the data we leave in our wake to make assumptions about who we are and how we will behave. Then they push us to behave more consistently with what they have determined to be our statistically most probable selves. They want us to be true to our profiles.

Everything we do in our highly connected reality is translated into data and stored for comparison and analysis. This includes not only which websites we visit, purchases we make, and photos we click on, but also real-world behaviors such as our driving styles and physical movements as tracked by mapping apps and GPS. Our smart thermostats and refrigerators all feed data into our profiles.

Most people worry about what specific information companies may record about us: we don't want anyone to know the content of our emails, what we look at for kicks, or what sorts of drugs we take. That's the province of crude web retailers who follow us with ads for things we've already bought. Algorithms don't care about any of that. The way they make their assessments of who we are and how to manipulate us has more to do with all the meaningless metadata they collect, compile, and compare.

For instance, Joe may travel twelve miles to work, look at his text messages approximately every sixteen minutes, purchase fat-free cookies and watch a particular TV program two days after it airs. The algorithm doesn't care about any of the specifics, nor does it attempt to make any logical conclusions about what kind of person Joe may be. All the algorithm cares about is whether this data allows it to put Joe in a statistical bucket along with other people like him, and if people in that bucket are likely to exhibit any similar behaviors in the future.

By crunching all these numbers and making constant comparisons between what we've done and what we do next, big-data algorithms can predict our behaviors with startling accuracy. Social media sites use the data they've collected about us to determine, with about 80 percent accuracy, who is about to get divorced, who is coming down with the flu, who is pregnant, and who may consider a change in sexual orientation—before we know ourselves.

Once algorithms have determined that Mary is, say, 80 percent likely to go on a diet in the next three weeks, they will fill her feeds with messages and news content about dieting: "Feeling fat?" Some of these messages are targeted marketing, paid for by the site's various advertisers. But the purpose of the messaging isn't just to sell any particular advertiser's products. The deeper objective is to get users to behave more consistently with their profiles and the consumer segment to which they've been assigned.

The social media platform wants to increase the probability Mary will go on a diet from 80 percent to 90 percent. That's why Mary's feeds fill up with all those targeted messages. The better it does at making Mary conform to her algorithmically determined destiny, the more the platform can boast both its predictive accuracy and its ability to induce behavior change.

Algorithms use our past behavior to lump us into statistical groups and then limit the range of choices we make moving forward. If 80 percent of people in a particular big-data segment are already planning to go on a diet or get divorced, that's fine. But what of the other 20 percent? What were they going to do instead? What sorts of anomalous behavior, new ideas, or novel solutions were they going to come up with before they were persuaded to fall in line?

In many human enterprises, there's a tendency toward the Pareto principle, or what's become known as the 80/20 rule: 80 percent of people will behave rather passively, like consumers, but 20 percent of people will behave more actively or creatively. For example, 80 percent of people watching videos online do only that; 20 percent of them make comments or post their own.

While 80 percent of kids play games as they were intended to be played, 20 percent of kids modify them or create their own levels. The people in the 20 percent open up new possibilities.

We are using algorithms to eliminate that 20 percent: the anomalous behaviors that keep people unpredictable, weird, and diverse. And the less variation among us—the less variety of strategies and tactics—the less resilient and sustainable we are as a species. Survivability aside, we're also less interesting, less colorful, and less human. Our irregular edges are being filed off.

In a stark reversal of figure and ground, we develop computer algorithms that constantly advance in order to make human beings more predictable and machinelike.

35.

Safer than the real world, where we are judged and our actions have consequences, virtual social spaces were assumed to encourage experimentation, role-playing, and unlikely relationships. Luckily for those depending on our alienation for profits, digital media doesn't really connect people that well, even when it's designed to do so. We cannot truly relate to other people online—at least not in a way that the body and brain recognize as real.

As neuroscientists have now established, human beings require input from organic, three-dimensional space in order to establish trusting relationships or maintain peace of mind. We remember things better when we can relate them to their physical locations, such as when we study from a book instead of a digital file.

The human nervous system calibrates itself over time based on the input we receive from the real world. A baby learns how to fall asleep by lying next to its mother and mirroring her nervous system. An anxious person gets calm after a walk in the woods. We come to trust another person by looking into their eyes and establishing rapport. We feel connected to a group when we breathe in unison.

Digital is just not good enough to fool the brain and body into these same states. Sure, it's close. Digital recordings have no "noise floor"—meaning no background hiss at all. But that's not the same as organic fidelity. A record album was an object in itself. We may have heard the clicks and scratches of an LP, but that allowed the brain and body to calibrate itself to the event occurring in the room—the vinyl record being played. The playback of a digital recording is less a real-world event than the projection of a symbolic event—a mathematical one—into the air. We have no reference for this.

Nor do we have an organic reference for a cellphone call or video conversation. We may look into the eyes of our conversation partner on a screen, but we can't see if their pupils are getting larger or smaller. Perhaps we can just make out their breathing rate and subconsciously pace ourselves to establish rapport—but it doesn't quite work. We're not getting much more information than we do from text, even though we're seeing someone's face or hearing their voice. This confuses us.

All the methods technologists use to increase the apparent fidelity of these exchanges are, themselves, fake—more noise than signal. The MP3 algorithm used for compressing music files, to take just one example, is not designed to represent the music accurately; it's designed to fool the brain into *believing* it is hearing music accurately. It creates some of the sensations we associate with bass or treble without using up valuable bandwidth to re-create the actual sounds. Through earbuds, the simulation is convincing. When played through stereo speakers, the missing information becomes apparent—less to the ears than to the whole body, which is expecting to absorb vibrations consistent with the music tones being simulated. But they're just not there. When we communicate through algorithmic compression, we are seeing what looks like a clearer image, or listening to what sounds like a more accurate voice—when they are anything but.

Audio engineers who care about fidelity try to restore it in other ways. The most common tactic to establish a wider dynamic range is to slow things down. One person says something, and the computer records it and stores it in a buffer before sending it to the receiver. The digital transmissions are paused for a fraction of a second so that the computer can catch up and put it all together. But that means you hear something a perceptible moment after I've said it. This addition of "latency" necessarily changes the timing of a conversation, making it impossible to establish a normal or reassuring rhythm of responses.

Human beings rely on the organic world to maintain our prosocial attitudes and behaviors. Online relationships are to real ones like internet pornography is to making love. The artificial experience not only pales in comparison to the organic one, but degrades our understanding of human connection. Our relationships become about metrics, judgments, and power—the likes and follows of a digital economy, not the resonance and cohesion of a social ecology.

All those painstakingly evolved mechanisms for social connection—for playing as a team—fail in the digital environment. But because mediated exchanges are new on the evolutionary timescale, we don't have a way to understand what is going on. We know that the communication is false, but we don't have a species experience of inaccurate, lifeless, or delayed

DOUGLAS RUSHKOFF

media transmissions through which to comprehend the loss of organic signal.

Instead of blaming the medium, we blame the other party. We read the situation as a social failure, and become distrustful of the person instead of the platform. Team Human breaks down.

36.

While Team Human may be compromised in the digital environment, team algorithm is empowered.

As our more resonant communication pathways fail us, it becomes harder to check in with one another, operate in a coordinated fashion, and express or even experience empathy. We lose all the self-reinforcing feedback loops of rapport: the mirror neurons and oxytocin that reward us for socializing. Surprisingly, the inability to establish trust in digital environments doesn't deter us from using them, but spurs *more* consumption of digital media. We become addicted to digital media precisely *because* we are so desperate to make sense of the neuromechanical experience we're having there. We are compelled to figure it out, calibrate our sensory systems, and forge hightouch relationships in a landscape that won't permit any of these things. We instead become highly individuated, alienated, and suspicious of one another.

Engagement through digital media is just a new way of being alone. Except we're not really alone out there—the space is inhabited by the algorithms and bots that seek to draw us into purchases, entertainment, and behaviors that benefit the companies that have programmed them. They outnumber us, like "non-player characters" in a video game. We are as likely, or more likely, to be engaging with a bot on the internet than we are to be engaging with another person. And the experience is likely to feel more rewarding as well.

Unlike the humans, the AIs online are increasingly connected to one another. Companies regularly and instantaneously sell data to one another. This is how products you may have looked at on one website magically show up as advertisements on the next. And that's just a primitive, obvious example of what's going on behind the scenes. The AIs are in constant communication, sharing with one another what they have learned by interacting with us. They are networked and learning.

The internet of things, or IOT as its proponents like to call it, is a name for the physical objects in this tremendous network of chips and algorithms seeking to understand and manipulate us. While a networked thermostat or baby monitor may have certain advantages for the consumer, its primary value is for the network to learn about our behaviors or simply extract data from them in order to place us in ever more granular statistical categories.

The algorithms directing these bots and chips patiently try one technique after another to manipulate our behavior until they get the results they have been programmed to deliver. These techniques haven't all been prewritten by coders. Rather, the algorithms randomly try new combinations of colors, pitches, tones, and phraseology until one works. They then share this information with the other bots on the network for them to try on other humans. Each one of us is not just up against whichever algorithm is attempting to control us, but up against them all.

If plants bind energy, animals bind space, and humans bind time, then what do networked algorithms bind? They bind *us*. On the internet of things, we the *people* are the things.

Human ideals such as autonomy, social contact, and learning are again written out of the equation, as the algorithms' programming steers everyone and everything toward instrumental ends. While human beings in a digital environment become more like machines, entities composed of digital materials—the algorithms—become more like living entities. They act as if they are our evolutionary successors.

No wonder we ape their behavior.

MECHANO-
MORPHISM

37.

When autonomous technologies appear to be calling all the shots, it's only logical for humans to conclude that if we can't beat them, we may as well join them. Whenever people are captivated—be they excited or enslaved—by a new technology, it becomes their new role model, too.

In the Industrial Age, as mechanical clocks dictated human time and factory machines outpaced human workers, we began to think of ourselves in very mechanical terms. We described ourselves as living in a "clockwork universe," in which the human body was one of the machines. Our language slowly became invested with mechanical metaphors: we needed to grease the wheels, crank up the business, dig deeper, or turn a company into a well-oiled machine. Even everyday phrases, such as "fueling up" for eating lunch or "he has a screw loose" for thinking illogically, conveyed the acceptance of humans as mechanical devices.

As a society, we took on the machine's values of efficiency, productivity, and power as our own. We sought to operate faster, with higher outputs and greater uniformity.

In the digital age, we think of our world as computational. Everything is data, and humans are processors. That logic does not *compute*. She *multitasks* so well she's capable of *interfacing* with more than one person in her *network* at a time. How about *leveling up* with some new *life hacks*?

The language alone suggests a new way for human beings to function in the digital media environment. Projecting human qualities onto machines—like seeing a car grille as a face or talking to a smartphone AI like a person—is called anthropomorphism. But this is the opposite: we are projecting machine qualities onto humans. Seeing a human being as a machine or computer is called mechanomorphism. It's not just treating machines as living humans; it's treating humans as machines.

38.

Having accepted our roles as processors in an information age, we strive to function as the very best computers we can be.

We multitask, assuming that—like our computers—we can do more than one thing at a time. Study after study has shown that human beings cannot multitask. When we try to do more than one thing at a time, we invariably get less done, less accurately, with less depth and less understanding. This is true even when we believe we have accomplished more. That's because, unlike computers, human beings do not have parallel processors. We have a single, holistic brain with two complementary hemispheres.

Computers have several sections of memory, working separately but in parallel. When a computer chip gets a problem, it breaks down the problem into steps and distributes those steps to its processors. Each processor produces an answer, and the answers are then reassembled. Human beings can't do that. We can imitate this process by switching really fast between one task and another—such as driving a car and sending a text message—but we can't actually do both simultaneously. We can only pretend, and often at our peril.

Drone pilots, for just one example, who monitor and neutralize people by remote control from thousands of miles away, experience higher rates of post-traumatic stress disorder than "real" pilots. This was unexpected by the military, which feared that remote bombing might desensitize cyber-pilots to killing.

DOUGLAS RUSHKOFF

One explanation for their higher rates of distress is that, unlike regular pilots, drone pilots often observe their targets for weeks before killing them. But the stress rates remain disproportionately high even for missions in which the pilots had no prior contact with the victims.

The more likely reason for the psychic damage is that the soldiers are trying to exist in more than one location at a time. They are in a facility in, say, Nevada, operating a lethal weapon system deployed on the other side of the planet. After dropping ordnance and killing a few dozen people, the pilots don't land their planes, climb out, and return to the mess hall or debrief with their fellow pilots. They simply log out, get into their cars, and drive home to the suburbs for dinner with their families. It's like being two different people in different places in the same day.

Except none of us is two people or can be in more than one place. Unlike a computer program, which can be copied and run from several different machines simultaneously, human beings only have one "instance" of themselves running at a time.

We may want to be like the machines of our era, but we can never be as good at being digital devices as the digital devices themselves. This is a good thing, and maybe the only way to remember that by aspiring to imitate our machines, we leave something even more important behind: our humanity.

39.

A media environment is the behavior, landscape, metaphors, and values that are engendered by a particular medium. The invention of text encouraged written history, contracts, the Bible, and monotheism. The clock tower in medieval Europe supported the hourly wages and time-is-money ethos of the

Industrial Age. It's another way that our prevailing technologies serve as role models for human attitudes and behaviors.

Particular media environments promote particular sorts of societies. The invention of the printing press changed our relationship to text and ideas by creating a sense of uniformity, while encouraging generic production and wide distribution. The television environment aided corporations in their effort to make America into a consumer culture, by helping people visualize the role of new products in their lives. Likewise, the internet platforms most of us use are not merely products but entire environments inhabited by millions of people, businesses, and bots. They constitute a new public square, Main Street, and fourth estate. By coming to recognize the embedded biases, or "affordances," of today's digital media technologies, we enable ourselves to leverage their benefits, and compensate for their deficits, rather than unconsciously conforming to their influences.

For example, the smartphone is more than just that device in our pocket. Along with all the other smartphones, it creates an environment: a world where anyone can reach us at any time, where people walk down public sidewalks in private bubbles, and where our movements are tracked by GPS and stored in marketing and government databases for future analysis. These environmental factors, in turn, promote particular states of mind, such as paranoia about being tracked, a constant state of distraction, and fear of missing out.

The digital media environment also impacts us collectively, as an economy, a society, and even as a planet. The breathtaking pace at which a digital company can reach "scale" has changed investors' expectations of what a stock's chart should look like, as well as how a CEO should surrender the long-term health of a company to the short-term growth of shares. The internet's emphasis on metrics and quantity over depth and quality has engendered a society that values celebrity, sensationalism, and numeric measures of success. The digital media environment expresses itself in the physical environment as well; the production, use, and disposal of digital technologies depletes scarce resources, expends massive amounts of energy, and pollutes vast regions of the planet.

Before we surrender to the notion that we live in a world entirely determined by our media, we must remember that the influence goes both ways: each of these media was itself the product of the society that invented it. The invention of writing made many much harder society harder to manage at scale. But writing may have itself arisen out of the preexisting need that tyrants of the period had to manage their hordes of slaves. We humans are partners with our media, both in their invention and in the ways we choose to respond to their influence.

Knowing the particular impacts of a media environment on our behaviors doesn't excuse our complicity, but it helps us understand what we're up against—which way things are tilted. This enables us to combat their effects, as well as the darker aspects of our own nature that they provoke.

40.

All media environments have better and worse things about them. Television helped us think of the planet as one big organism, but also promoted consumerism and neoliberalism. The internet helps decentralize power and thought leadership, but also atomizes and isolates us from one another. Neither environment is necessarily better, but each new one requires a different response.

The hardest part of learning to respond to a new medium

is being able to see its effects in their own right, rather than through the lens of the prior era. When the internet emerged, most of us were living under the influence of the television media environment. The TV era was about globalism, international cooperation, and the open society. Television let people see for the first time what was happening in other places, often broadcast live, just as it happened. The whole world witnessed the same wars, floods, and revolutions together, in real time. Even 9/11 was a simultaneously experienced, global event, leading to almost universally expressed empathy. Television connected us as a planet.

As if invented to continue this trend, the internet was supposed to break down those last boundaries between what are essentially synthetic nation-states and herald a new, humanistic, global community of peers. National governments were declared extinct, and a new self-organizing network of humans was on the way. Yet the internet age has actually heralded the opposite result. We are not advancing toward some new, totally inclusive global society, but retreating back to nativism. Instead of celebrating more racial intermingling, we find many yearning for a fictional past when—people like to think—our races were distinct, and all was well.

At the height of the television media era, an American president could broadcast a speech in front of the Brandenburg Gate in Berlin and demand that Russia "tear down this wall." No more. Politicians of the digital media environment pull out of global trade blocs, and demand the construction of walls to enforce their countries' borders.

This is very different from the television environment, which engendered a "big blue marble" melting pot, hands-across-the-world, International Space Station, cooperative international-

DOUGLAS RUSHKOFF

ism that still characterizes our interpretations of geopolitics. Those of us flummoxed by the resurgence of nationalist, regressively antiglobal sentiments are mistakenly interpreting politics through that now-obsolete television screen.

The first protests of the digital media landscape, such as those against the World Trade Organization in Seattle, made no sense to the TV news. They seemed to be an incoherent amalgamation of disparate causes: environmentalists, labor activists, even anti-Zionists. What unified them, however—more than their ability to organize collectively on the internet—was their shared antiglobalism. The protesters had come to believe that the only entities capable of acting on the global level were ones too big for human beings to control.

The breakdown of European cohesion extends this sentiment. The European Union was a product of the television environment: open trade, one currency, free flow of people across boundaries, and the reduction of national identities to cuisine and soccer teams. The transition to a digital media environment is making people less tolerant of this dissolution of boundaries. Am I Croatian or Serbian? Kurd or Sunni? Greek or European?

The yearning for boundaries emerges from a digital media environment that emphasizes distinction. Everything is discrete. Analog media such as radio and television were continuous, like the sound on a vinyl record. Digital media, by contrast, are made up of many discrete samples. Likewise, digital networks break up our messages into tiny packets and reassemble them on the other end. Computer programs all boil down to a series of 1's and 0's, on or off. This logic trickles up to the platforms and apps we use. Everything is a choice—from font size to the place on a "snap-to" grid. It's either 12 point or 13 point, positioned here or there. Did you send the email or not? There are no in-betweens.

A society functioning on these platforms tends toward similarly discrete formulations. Like or unlike? Black or white? Rich or poor? Agree or disagree? In a self-reinforcing feedback loop, each choice we make is noticed and acted upon by the algorithms personalizing our news feeds, further isolating each one of us in our own ideological filter bubble. The internet reinforces its core element: the binary. It makes us take sides.

41.

Digital media push us apart, but they also seem to push us backward. Something about this landscape has encouraged the regressive sentiments of the populist, nationalist, and nativist movements characterizing our time. These sentiments grow in an ecosystem fed by the other main bias of digital media: memory.

Memory is what computers were invented for in the first place. In 1945, when Vannevar Bush imagined the "memex," on which computers were based, he described it as a digital filing cabinet—an external memory. And even though they can now accomplish much more than data retrieval, everything computers do—all of their functions—simply involves moving things from one part of their memory to another. Computer chips, USB sticks, and cloud servers are all just kinds of memory. Meanwhile, unnerving revelations about cybersecurity and surveillance continually remind us that everything we do online is stored in memory. Whatever you said or did on your favorite social network or search engine is in an archive, timeline, or server somewhere, waiting to be retrieved by someone, someday.

The unchecked exaltation of memory in a digital media environment, combined with a bias toward discrete boundaries, yields the political climate we're witnessing: Brexiteers justifying

isolation as a confirmation of distinctly British values and the return to a nationalist era when the rest of Europe was across the Channel; America's alt-right recalling a clearly redlined past when being white and American meant enjoying a segregated neighborhood, a sense of superiority, and a guaranteed place in the middle class. Immigrants were fellow Germans, Irish, and Italians—not nonwhites, "foreigners," refugees, or terrorists leaking illegally across permeable national boundaries.

To be sure, globalism had some genuinely devastating effects on many of those who are now pushing back. Wealth disparity is at an all-time high, as global trade and transnational banks dwarf the mitigating effects of local and national economic activity. But the way people are responding to this pressure, in the West anyway, is digital in spirit.

The digital media environment is hardly isolated, however, to Western, developed economies. Across the world, we see digital sensibilities promoting similarly impulsive, nativist behaviors. The rise of genocidal fervor against the Rohingya in Myanmar has correlated directly with the rise of social media use there. Similar tensions are rising in India, Malaysia, and Sudan, all fueled by digital media's ability to provoke emotions, spread unchecked facts, and trigger false memories of a better, purer past.

Those of us who want to preserve the prosocial, one-world vision of the TV media environment, or the reflective intellectualism of the print era, are the ones who must stop looking back. If we're going to promote connection and tolerance, we'll have to do it in a way that recognizes the biases of the digital media environment in which we are actually living, and then encourages human intervention in these otherwise automated processes.

Resistance is futile. The word "resistance" itself is a relic of the electronic age, where a resistor on a circuit board could attenuate the current passing through it. There is no resistance in a digital environment—only on or off. Anything in-between is relegated to one or the other, anyway. We can't attenuate the digital. There is no volume knob. There are no knobs at all; there are only switches.

In a digital media environment there is no resistance, only opposition.

42.

It's hard for human beings to oppose the dominance of digital technology when we are becoming so highly digital ourselves. Whether by fetish or mere habit, we begin acting in ways that accommodate or imitate our machines, remaking our world and, eventually, ourselves in their image.

For instance, the manufacturers of autonomous vehicles are encouraging cities to make their streets and signals more compatible with the navigation and sensor systems of the robotic cars, changing our environment to accommodate the needs of the robots with which we will be sharing the streets, sidewalks, and, presumably, air space. This isn't so bad in itself, but if history is any guide, remaking the physical world to accommodate a new technology—such as the automobile—favors the companies selling the technologies more than the people living alongside them. Highways divided neighborhoods, particularly when they reinforced racial and class divisions. Those who couldn't adapt to crosswalks and traffic signals were labeled "jaywalkers" and ridiculed in advertisements.

Today, we are in the process of making our physical, social, and media environments more friendly to digital operants—

often ignoring the impact on real people. New studies on the health effects of cellphones and WiFi are ignored or buried as fast as they can be produced, and our schools and municipalities become more heavily and irreversibly invested in wireless networks, smart boards, and other computer-based learning.

As if to celebrate their commitment to digital values, some school principals encourage teachers to post on social media throughout the day. Education becomes fodder for the feeds, and subordinate to its promotion on the net. Lessons become valued for their ability to be photographed and posted. And teachers model for students the addictive behaviors these platforms are designed to instill.

Even the aesthetics of a culture eventually adapt to those suggested by its dominant technologies. Digital technology may have reduced music to MP3 algorithms that convey only the idea of a sound instead of its essence, but the digital environment has also reduced performers to autotuned commodities. The market has exercised such control over music since the advent of recording studios and their appropriately named "control rooms," but today we see that dynamic amplified by new technology. The producer behind the glass is in charge of the mixing board and the performers—who are relegated to digitally separated tracks. "Isolated" from one another, to use music recording terminology, the performers lose the camaraderie and rapport of live group performance, and instead sync themselves to a computer-generated metronomic beat. If they fall off the rhythm, even infinitesimally, there's an autocorrect feature in the recording technology to force their performance back to inhuman perfection.

It sounds "better"—or at least more accurate to the pitch and rhythm. But what is the perfect note and pace, really? The one mathematically closest to the predetermined frequency? Any decent violinist will tell you that an E-flat and a D-sharp may look like the same note, but are actually subtly different depending on the key of the song and the notes around it. Different musicians might even interpret the note differently depending on its context, or slide up to the note—intentionally emphasizing the effort required—or slide down after hitting it, as if diminishing in conviction.

Musicians also vary ever so slightly from the exact rhythm of a song to create many effects, or even just as a true expression of their approach to music and life. Ringo Starr, the Beatles drummer, famously lagged ever so slightly behind the beat—as if to express a laziness or "falling down the stairs" quality of playing. Ringo's delay is human, and so close to the "normal" beat of the song that it would be immediately corrected by production technologies biased toward making humans sound just as "good" as computers.

Our mechanomorphic culture is embracing a digital aesthetic that irons out anything uniquely human. Any quirks of voice or intonation—gravel, wobble, air, or slide—are reinterpreted as imperfections. The ideal is perfect fidelity—not to the human organisms actually performing the music, but to the mathematics used to denote the score. We forget that those notations are an approximation of music, a compromised way of documenting an embodied expression of human emotion and artistry as a system of symbols so that it can be re-created by someone else.

The figure and ground are reversed when the human performance is seen as an impediment to the pure data, rather than a way of connecting people on both perceived and unconscious levels. The noises that emanate from the human beings or their instruments are not expressions of autonomy but samples to be manipulated: raw material for digital processing or labor to be extracted and repackaged.

Human interpretation no longer matters, and any artifacts of our participation are erased. We may as well be machines.

43.

The highest ideal in an unrestrained digital environment is to transcend one's humanity altogether. It's a straightforward logic: if humanity is a purely mechanistic affair, explicable entirely in the language of data processing, then what's the difference whether human beings or computers are doing that processing?

Transhumanists hope to transcend or at least improve upon biological existence. Some want to use technology to live forever, others to perform better, and still others to exit the body and find a better home for consciousness. Following through on mechanomorphism, transhumanism holds that people can be upgraded just like machines. By blurring the line between what we think of as biological and technological, transhumanists hope to ease our eventual, inevitable transition to life on a silicon chip.

As proponents will argue, our journey into a transhumanist state of existence has already started, with dentures, hearing aids, and artificial hearts—all synthetic extensions that change what it means to be human. Would we call contact lenses anti-human? Of course not. So why scoff at brain implants that enable us to speak new languages? Or why reject an offer to clone one's consciousness and upload it to a server?

It's a compelling thought: exchanging one's humanity for immortality. Piece by piece, or maybe all at once, we become what transhumanists argue is the next stage of evolution: some hybrid of person and machine, or maybe an entirely digital species. Along the way, though, we take on an increasingly mechanistic understanding of our personhood.

Ironically, transhumanism is less about embracing the future than fixing the human experience as it is today. Medical and life extension interventions seek only to preserve the person who is alive right now. Cryonics seeks to freeze the human form in its current state in order to be reanimated in the future. Uploading one's mind simply transfers the human brain, or a perfect clone of it, as is, into a more durable substrate.

In some ways, transhumanism is a reactionary response to the sorts of changes inherent in nature, a defiant assertion of the individual against its own impermanence. The cycles of life are understood not as opportunities to learn or let go, but as inconveniences to ignore or overcome. We do not have to live with the consequences of our own actions. There's an app for that.

44.

Self-improvement of the transhumanist sort requires that we adopt an entirely functional understanding of who and what we are: all of our abilities can be improved upon and all of our parts are replaceable. Upgradable.

The quirks that make us human are interpreted, instead, as faults that impede our productivity and progress. Embracing those flaws, as humanists tend to do, is judged by the transhumanists as a form of nostalgia, and a dangerously romantic misinterpretation of our savage past as a purer state of being. Nature and biology are not mysteries to embrace but limits to transcend.

This transhumanist mindset is, in fact, taking hold. We can

DOUGLAS RUSHKOFF

see it in the way we bring digital technologies closer and closer. The screen is encroaching on the eye, from TVs to computer monitors to phone screens to smart watches to VR goggles to tiny LEDs that project images onto the retina to neural implants that communicate directly with the optic nerve.

With each leap in human–machine intimacy, resolution increases, and our utility value is improved along some measurable metric. This is the mindset encouraged by wristbands that count our heartbeats and footsteps under the pretense of improved health or life extension. Health, happiness, and humanity itself are all reducible to data points and subject to optimization.

We are all just numbers: the quantified self.

Like a music recording that can be reduced to code and stored in a file, the quantified human can also be reduced to bits, replicated infinitely, uploaded to the cloud, or installed in a robot. But only the metrics we choose to follow are recorded and translated. Those we don't value, or don't even know about, are discarded in the new model.

Improved longevity, supercognition, or military prowess all sound promising until we consider what is being left behind, whose values are being expressed (and whose aren't), and how these choices change the larger systems of which we are a part. Just as life-saving antibiotics also strengthen bacteria and weaken the collective immune system, or as steroids improve short-term performance at the expense of long-term health, there are trade-offs. We—or the companies selling the improvements—are actively picking which features of humanity to enhance and which to suppress or ignore. In amplifying an individual's brainpower, we may inadvertently disable some of their ability to connect with others or achieve organismic

resonance. What of racial diversity, gender fluidity, sexual orientation or body type? The human traits that are not favored by the market will surely be abandoned.

Could that happen to a civilization as supposedly enlightened as our own? Our track record suggests it will.

The internet's tremendous social and intellectual potential was surrendered to short-term market priorities, turning a human-centered medium into a platform for manipulation, surveillance, and extraction. The more we see the human being as a technology to be enhanced, the greater the danger of applying this same market ethos to people, and extending our utility value at the expense of others. Life extension becomes the last-ditch attempt of the market to increase our available timeline as consumers—and consumers willing to spend anything for that extra few years of longevity.

Sure, many of us would accept a digital implant if it really worked as advertised. Who wouldn't want some painless enhancements, free of side effects? Or the choice of when or whether to die? Besides, participation in the ever-changing economy requires some acquiescence to technology, from streetcars and eyeglasses to elevators and computers.

But our technological investments come with strings. The deals are not honest; they've never been. Companies change their user agreements. Or they sell the printers at a loss and then overcharge us for the ink cartridges. The technologies we are currently bringing into our lives turn us into always-on customers—more like subscribers than purchasers, never really owning or fully controlling what we've bought. Operating system upgrades render our hardware obsolete, forcing us to buy new equipment. How long until that chip in my brain, and the neurons that grew around it, are obsolete?

It's not that wanting to improve ourselves, even with seemingly invasive technology, is so wrong. It's that we humans should be making active choices about what it is we want to do to ourselves, rather than letting the machines, or the markets propelling them, decide for us.

ECONOMICS

45.

Technology is not driving itself. It doesn't *want* anything. Rather, there is a market expressing itself through technology— an operating system beneath our various computer interfaces and platforms that is often unrecognized by the developers themselves. This operating system is called capitalism, and it drives the antihuman agenda in our society at least as much as any technology.

Commerce is not the problem. People and businesses can transact in ways that make everyone more prosperous. If anything, capitalism as it's currently being executed is the enemy of commerce, extracting value from marketplaces and delivering it to remote shareholders. The very purpose of the capitalist operating system is to prevent widespread prosperity.

What we now think of as capitalism was born in the late Middle Ages, in the midst of a period of organic economic growth. Soldiers had just returned from the Crusades, having opened up new trade routes and bringing back innovations from foreign lands. One of them, from the Moorish bazaar, was the concept of "market money."

Until this point, European markets operated mostly through barter, the direct exchange of goods. Gold coins, like the florin, were just too scarce and valuable to be spent on bread. Anyone who had gold—and most peasants did not—hoarded it. Market money let regular people sell their goods to one another. It was often issued in the morning, like chips at the beginning of a poker game, and then cashed in at the close of trading. Each unit of currency might represent a loaf of bread or a head of lettuce, and would be used as credit by the seller of those items as a

way of priming the pump for the day's trade. So the baker could go out early and buy the things he needed, using coupons good for a loaf of bread. Those coupons would slowly make their way back to the baker, who would exchange them for loaves of bread.

The Moors also invented grain receipts. A farmer could bring a hundred pounds of grain to the grain store and leave with a receipt. It would be perforated into ten-pound increments, so that the farmer could tear off a portion and spend it to buy what he needed. The interesting thing about this form of money is that it lost value over time. The grain store had to be paid, and some grain was lost to spoilage. So the money itself was biased toward spending. Who would hold onto money that was going to be worth less next month?

This was an economy geared for the velocity of money, not the hoarding of capital. It distributed wealth so well that many former peasants rose to become the new merchant middle class. They worked for themselves, fewer days per week, with greater profits, and in better health than Europeans had ever enjoyed (or would enjoy again for many centuries).

The aristocracy disliked this egalitarian development. As the peasants became self-sufficient, feudal lords lost their ability to extract value from them. These wealthy families hadn't created value in centuries, and so they needed to change the rules of business to stem the rising tide of wealth as well as their own decline.

They came up with two main innovations. The first, the chartered monopoly, made it illegal for anyone to do business in a sector without an official charter from the king. This meant that if you weren't the king's selected shoemaker or vintner, you had to close your business and become an employee of someone who was. The American Revolution was chiefly a response to such monopoly control by the British East India Company. Colonists were free to grow cotton but forbidden from turn-

ing it into to fabric or selling it to anyone but the company, at exploitative prices. The company transported the cotton back to England, where it was made into fabric, then shipped it back to America and sold it to the colonists. The monopoly charter was the progenitor of the modern corporation,.

The other main innovation was central currency. Market money was declared illegal; its use was punishable by death. People who wanted to transact had to borrow money from the central treasury, at interest. This was a way for the aristocracy, who had money, to make money simply by lending it. Money, which had been a utility to promote the exchange of goods, became instead a way of extracting value from commerce. The local markets collapsed.

The only ones who continued to borrow were the large, chartered monopolies. Of course, if companies had to pay back more money than they borrowed, they had to get the additional funds somewhere. This meant that the economy had to grow. So the chartered corporations set out to conquer new lands, exploit their resources, and enslave their peoples.

That growth mandate remains with us today. Corporations must grow in order to pay back their investors. The companies themselves are just the conduits through which the operating system of central currency can execute its extraction routines. With each new round of growth, more money and value is delivered up from the real world of people and resources to those who have the monopoly on capital. That's why it's called capitalism.

46.

If central currency can be thought of as the operating system of our economy, corporations are the software that runs on top of it. They are the true natives of capitalism, which is why they are

so much more at home in this environment than we humans. In perhaps the most spectacular reversal of figure and ground we've yet witnessed, corporations have been winning court cases that give them the rights of human beings—from personhood and property to free speech and religious convictions—while human beings now strive to brand themselves in the style of corporations.

But corporations are not people. They are abstract, and can scale up infinitely to meet the demands of the debt-based economy. People can only work so hard or consume so much before we reach our limits. We are still part of the organic world, and subject to the laws of nature. Corporations know no such bounds, making them an awful lot like the digital technologies they are developing and inhabiting.

The pioneering philosopher of the political economy, Adam Smith, was well aware of the abstract nature of corporations—particularly large ones—and stressed that regulations would be necessary to keep them from destroying the marketplace. He argued that there are three factors of production, which must all be recognized as equally important: the land, on which we grow the crops or extract the resources; the labor, who till the soil or manufacture the goods; and, finally, the capital—either the money invested or the tools and machines purchased. He saw that in an abstract, growth-based economy, the priorities of the capital could quickly overtake the other two, and that this, in turn, would begin to favor the largest corporate players over the local, human-scaled enterprises that fuel any real economy.

Unlike land and humans, which are fixed, capital can keep growing. It has to, because a growth-based economy always requires more money. And capital accomplishes this miracle growth by continually abstracting itself. Investors who don't want to wait three months for a stock to increase in value can use a derivative—an abstraction—to purchase the future stock

now. If that's not enough temporal compression, they can purchase a derivative of that derivative, and so on. Today, derivatives trading far outpaces trading of real stocks—so much so that the New York Stock Exchange was actually purchased by its derivatives exchange in 2013. The stock exchange, itself an abstraction of the real marketplace of goods and services, was purchased by its own abstraction.

As more of the real world becomes subjected to the logic of capital, the things, people, and places on which we depend become asset classes. Homes become too expensive for people to buy because they're a real estate investment for investors, hedge funds and other nonhumans. Those who do manage to purchase homes soon realize that they're only providing fodder for the mortgage industry, whose mortgages, in turn, are bundled into a yet more abstracted asset class of their own.

People are at best an asset to be exploited, and at worst a cost to be endured. Everything is optimized for capital, until it runs out of world to consume.

47.

Growth was easy when there were new territories to conquer, resources to take, and people to exploit. Once those people and places started to push back, digital technology came to the rescue, providing virtual territory for capital's expansion. Unfortunately, while the internet can scale almost infinitely, the human time and attention that create the real value are limited.

Digital companies work the same way as their extractive forebears. When a big box store moves to a new neighborhood, it undercuts local businesses and eventually becomes the sole retailer and employer in the region. With its local monopoly, it can then raise prices while lowering wages, reduce labor to part-time status, and externalize the costs of healthcare and food stamps to the government. The net effect of the business on the community is extractive. The town becomes poorer, not richer. The corporation takes money out of the economy—out of the land and labor—and delivers it to its shareholders.

A digital business does the same thing, only faster. It picks an inefficiently run industry, like taxis or book publishing, and optimizes the system by cutting out most of the people who used to participate. So a taxi service platform charges drivers and passengers for a ride while externalizing the cost of the car, the roads, and the traffic to others. The bookselling website doesn't care if authors or publishers make a sustainable income; it uses its sole buyer or "monopsony" power to force both sides to accept less money for their labor. The initial monopoly can then expand to other industries, like retail, movies, or cloud services.

Such businesses end up destroying the marketplaces on which they initially depend. When the big box store does this, it simply closes one location and starts the process again in another. When a digital business does this, it pivots or expands from its original market to the next—say, from books to toys to all of retail, or from ride-sharing to restaurant delivery to autonomous vehicles—increasing the value of its real product, the stock shares, along the way.

The problem with this model, from a shareholder perspective, is that it eventually stops working. Even goosed by digital platforms, corporate returns on assets have been steadily declining for over seventy-five years. Corporations are still great at sucking all of the money out of a system, but they're awful at deploying those assets once they have them. Corporations are getting bigger but less profitable. They're just sitting on piles of unused money, and taking so much cash out of the system that central banks are forced to print more. This new money gets invested in banks that lend it to corporations, starting the cycle all over again.

Digital businesses are just software that converts real assets into abstract forms of shareholder value. Venture capitalists remain hopeful that they will invest in the next unicorn with a "hockey stick"–shaped growth trajectory, and then get out before the thing crashes. These businesses can't sustain themselves, because eventually the growth curve must flatten out.

The myth on which the techno-enthusiasts hang their hopes is that new innovations will continue to create new markets and more growth. For most of history, this has been true, sort of. Just when agriculture reached a plateau, we got the steam engine. When consumerism stalled, television emerged to create new demand. When web retail slowed its growth, we got data mining. When data as a commodity seemed to plateau, we got artificial intelligence, which needs massive supplies of data in order to learn.

Except in order to stoke and accelerate growth, new, paradigm-busting inventions like smartphones, robots, and drones must not only keep coming, but keep coming faster and faster. The math doesn't work: we are quickly approaching the moment when we will need a major, civilization-changing innovation to occur on a monthly or even weekly basis in order to support the rate of growth demanded by the underlying operating system. Such sustained exponential growth does not occur in the natural world, except maybe for cancer—and that growth ceases once the host has been consumed.

48.

Instead of bringing widespread prosperity, the digital economy has amplified the most extractive aspects of traditional capitalism. Connectivity may be the key to participation, but it also gives corporations more license and capacity to extract what little value people have left. Instead of retrieving the peer-to-peer marketplace, the digital economy exacerbates the division of wealth and paralyzes the social instincts for mutual aid that usually mitigate its effects.

Digital platforms amplify the power law dynamics that determine winners and losers. While digital music platforms make space for many more performers to sell their music, their architecture and recommendation engines end up promoting many fewer artists than a diverse ecosystem of record stores or FM radio did. One or two superstars get all the plays, and everyone else sells almost nothing.

It's the same across the board. While the net creates more access for artists and businesses of all kinds, it allows fewer than ever to make any money. The same phenomenon takes place on the stock market, where ultra-fast trading algorithms spur unprecedented momentum in certain shares, creating massive surpluses of capital in the biggest digital companies and sudden, disastrous collapses of their would-be competitors. Meanwhile, automation and extractive platforms combine to disadvantage anyone who still works for a living, turning what used to be lifelong careers into the temp jobs of a gig economy.

These frictionless, self-reinforcing loops create a "winner takes all" landscape that punishes the middle class, the small business, and the sustainable players. The only ones who can survive are artificially inflated companies, who use their ballooning share prices to purchase the also-rans. Scale is everything. This sensibility trickles down to all of us, making us feel like our careers and lives matter only if we've become famous, earned a million views, or done something, even something destructive, "at scale."

While digital business plans destroy a human-scaled economy, the digital businesses themselves compromise the human sensibilities required to dig ourselves out of this mess. The human beings running those enterprises are no less the psychic victims of their companies' practices than the rest of us, which is why it's so hard for them to envision a way out.

Well-meaning developers, who have come to recognize the disastrous impacts of their companies, seek to solve technology's problems with technological solutions. They see that social media algorithms are exacerbating wealth division and mental confusion, and resolve to tweak them not to do that at least not so badly. The technosolutionists never consider the possibility that some technologies themselves have intrinsic antihuman affordances. (Guns may not kill people, but they are more biased toward killing than, say, pillows, even though both can be used for that purpose.) Furthermore, they propose technosolutions that are radical in every way except in their refusal to challenge the underlying rule set of venture capitalism or the extreme wealth of those who are making the investments. Every technosolution must still be a profitable investment opportunity—otherwise, it is not considered a solution at all.

Even promising wealth redistribution ideas, such as universal basic income, are recontextualized by the technosolutionists as a way of keeping their companies going. In principle, the idea of a negative income tax for the poor, or a guaranteed minimum income for everyone, makes economic sense. But when we hear these ideas espoused by Silicon Valley's CEOs, it's usually in the context of keeping the extraction going. People have been sucked dry, so now the government should just print more money for them to spend. The argument merely reinforces the human obligation to keep consuming, or to keep working for an unlivable wage.

More countercultural solutions, such as bitcoin and the blockchain, are no less technosolutionist in spirit. The blockchain replaces the need for central authorities such as banks by letting everyone on a network authenticate their transactions with computer encryption. It may disintermediate exploitative financial institutions but it doesn't help rehumanize the economy, or reestablish the trust, cohesion, and ethos of mutual aid that was undermined by digital capitalism. It simply substitutes for trust in a different way: using the energy costs of blockchain mining as a security measure against counterfeiting or other false claims. (The computer power needed to create one bitcoin consumes at least as much electricity as the average American household burns through in two years.) Is this the fundamental fix we really need? A better ledger?

The problem the blockchain solves is the utilitarian one of better, faster accounting, and maybe an easier way to verify someone's identity online. That's why the banking industry has ultimately embraced it: the quicker to find us and drain our assets. Progressives, meanwhile, hope that the blockchain will be able to record and reward the unseen value people are creating as they go about their lives—as if all human activity were transactional and capable of being calculated by computer.

We must learn that technology's problems can't always be solved with more technology.

49.

Some of the more farsighted tech billionaires are already investing in plan B. Instead of undoing the damage, reforming their companies, or restoring the social compact, they're busy preparing for the apocalypse.

The CEO of a typical company in 1960 made about 20 times

as much as its average worker. Today, CEOs make 271 times the salary of the average worker. Sure, they would like to take less and share with their workers, but they don't know how to give up their wealth safely. As Thomas Jefferson once described the paradox of wanting to free his slaves but fearing their retribution if he did, it's like "holding a wolf by the ear." But why do you think his slaves were so angry in the first place?

Similarly, the perception of inequality is itself the main reason human beings treat one another less charitably. It's not the total amount of abundance in the system that promotes goodwill, but the sense that whatever is available is being distributed justly. The approximately five hundred families, who own 80 percent of the world's assets, are so worried about the impoverished classes staging an uprising—either now or after a disaster—that they feel they must continue to build up cash, land, supplies, and security.

They hire futurists and climatologists to develop strategies for different scenarios, and then purchase property in Vancouver, New Zealand, or Minneapolis—regions predicted to be least affected by rising sea levels, social unrest, or terror strikes. Others are investing in vast underground shelters, advanced security systems, and indoor hydroponics in order to withstand a siege from the unruly world. The most energetic billionaires are busy developing aerospace and terraforming technologies for an emergency escape to a planet as yet unspoiled by their own extractive investment practices.

These oligarchs deploy an "insulation equation" to determine how much of their fortunes they need to spend in order to protect themselves from the economic, social, and environmental damage that their business activities have caused. Even the new headquarters of the biggest Silicon Valley firms are built more like fortresses than corporate parks, micro-feudal empires turned inward on their own private forests and gardens and protected from the teeming masses outside.

Such expenditures, no matter how obscene, represent what investors consider a calculated "hedge" against bad times. They don't actually believe the zombies are at the gate; they just want some insurance against the worst-case scenario.

Of course, there's a better, more human way to calculate the insulation equation: instead of determining the investment required to insulate oneself from the world, we can look instead at how much of our time, energy, and money we need to invest in the world so that it doesn't become a place we need to insulate ourselves from in the first place.

50.

The economy needn't be a war; it can be a commons. To get there, we must retrieve our innate good will.

The commons is a conscious implementation of reciprocal altruism. Reciprocal altruists, whether human or ape, reward those who cooperate with others and punish those who defect. A commons works the same way. A resource such as a lake or a field, or a monetary system, is understood as a shared asset. The pastures of medieval England were treated as a commons. It wasn't a free-for-all, but a carefully negotiated and enforced system. People brought their flocks to graze in mutually agreed-upon schedules. Violation of the rules was punished, either with penalties or exclusion.

The commons is not a winner-takes-all economy, but an all-take-the-winnings economy. Shared ownership encourages shared responsibility, which in turn engenders a longer-term perspective on business practices. Nothing can be externalized to some "other" player, because everyone is part of the same trust, drinking from the same well.

If one's business activities hurt any other market participant, they undermine the integrity of the marketplace itself. For those entranced by the myth of capitalism, this can be hard to grasp. They're still stuck thinking of the economy as a two-column ledger, where every credit is someone's else's debit. This zero-sum mentality is an artifact of monopoly central currency. If money has to be borrowed into existence from a single, private treasury and paid back with interest, then this sad, competitive, scarcity model makes sense. I need to pay back more than I borrowed, so I need to get that extra money from someone else. That's the very premise of zero-sum. But that's not how an economy has to work.

The destructive power of debt-based finance is older than central currency—so old that even the Bible warns against it. It was Joseph who taught Pharaoh how to store grain in good times so that he would be able to dole it out in lean years. Those indentured to the pharaoh eventually became his slaves, and four hundred years passed before they figured out how to free themselves from captivity as well as this debtor's mindset. Even after they escaped, it took the Israelites a whole generation in the desert to learn not to hoard the manna that rained on them, but to share what came and trust that they would get more in the future.

If we act like there's a shortage, there will be a shortage.

51.

Advocates of the commons seek to optimize the economy for human beings, rather than the other way around.

One economic concept that grew out of the commons was called distributism. The idea, born in the 1800s, holds that instead of trying to redistribute the spoils of capitalism after the fact through heavy taxation, we should simply predistribute the means of production to the workers. In other words, workers should collectively own the tools and factories they use to create value. Today, we might call such an arrangement a co-op—and, from the current examples, cooperative businesses are giving even established US corporations a run for their money.

The same sorts of structures are being employed in digital businesses. In these "platform cooperatives," participants own the platform they're using, instead of working for a "platform monopoly" taxi app or giving away their life data to a social media app. A taxi app is not a complicated thing; it's just a dating app combined with a mapping app combined with a credit card app. The app doesn't deserve the lion's share of the revenue. Besides, if the drivers are going to be replaced by robots someday, anyway, at least they should own the company for which they've been doing the research and development. Similarly, a user-owned social media platform would allow participants to sell (or not sell) their own data, instead of having it extracted for free.

Another commons-derived idea, "subsidiarity," holds that a business should never grow for growth's sake. It should only grow as big as it needs to in order to accomplish its purpose. Then, instead of expanding to the next town or another industry, it should just let someone else replicate the model. Joe's pizzeria should sell to Joe's customers. If they need a pizzeria in the next town, Joe can share his recipe and let Samantha do it.

This is not bad business—especially if Joe likes making pizza. He gets to stay in the kitchen doing what he loves instead of becoming the administrator of a pizza chain. Samantha may develop a new technique that helps Joe; they can even federate and share resources. Besides, it's fun to have someone else to talk with about the pizza business. They can begin to develop their collaborative abilities instead of their competitive ones.

Bigger isn't necessarily better. Things in nature grow to a certain point and then stop. They become full-grown adults, forests, or coral reefs. This doesn't mean they're dead. If anything, it's the stability of adulthood that lets them become participating members of larger, mutually supportive networks.

If Joe has to grow his business bigger just in order to keep up with his rising rent and expenses, it's only because the underlying economy has been rigged to demand growth and promote scarcity. It is this artificially competitive landscape that convinces us we have no common interests.

52.

We know that nothing in nature can sustain an exponential rate of growth, but this doesn't stop many of our leading economists and scientists from perpetuating this myth. They cherry-pick evidence that supports the endless acceleration of our markets and our technologies, as if to confirm that growth-based corporate capitalism is keeping us on track for the next stage of human evolution.

To suggest we slow down, think, consider—or content ourselves with steady profits and incremental progress—is to cast oneself as an enemy of our civilization's necessary acceleration forward. By the market's logic, human intervention in the machine will only prevent it from growing us out of our current mess. In this read of the situation, corporations may be using extractive, scorched-earth tactics, but they are also our last best hope of solving the world's biggest problems, such as hunger and disease. Questioning the proliferation of patented, genetically modified seeds or an upgraded arsenal of pesticides just impedes the necessary progress. Adherents of this worldview say that it's already too late to go back. There are already too many people, too much damage, and too much dependence on energy. The only way out is through. Regulating a market just slows it down, preventing it from reaching the necessary level of turbulence for the "invisible hand" to do its work.

According to their curated history of humanity, whenever things look irredeemably awful, people come up with a new technology, unimaginable until then. They like to tell the story of the great horse manure crisis of 1894, when people in England and the United States were being overwhelmed by the manure produced by the horses they used for transportation. Luckily, according to this narrative, the automobile provided a safe, relatively clean alternative, and the streets were spared hip-deep manure. And just as the automobile saved us from the problems of horse-drawn carriages, a new technological innovation will arise to save us from automobiles.

The problem with the story is that it's not true. Horses were employed for commercial transport, but people rode in electric streetcars and disliked sharing the roads with the new, intrusive, privately owned vehicles. It took half a century of public relations, lobbying, and urban replanning to get people to drive automobiles. Plus, we now understand that if cars did make the streets cleaner in some respects, it was only by externalizing the costs of environmental damage and the bloody struggle to secure oil reserves.

Too many scientists—often funded by growth-obsessed corporations—exalt an entirely quantified understanding of social progress. They measure improvement as a function of life expectancy or reduction in the number of violent deaths. Those are great improvements on their own, but they give false cover for the crimes of modern capitalism—as if the relative peace and longevity enjoyed by some inhabitants of the West were proof of the superiority of its model and the unquestionable benefit of pursuing growth.

These arguments never acknowledge the outsourced slavery, toxic dumping, or geopolitical strife on which this same model depends. So while one can pluck a reassuring statistic to support the notion that the world has grown less violent—such as the decreasing probability of an American soldier dying on the battlefield—we also live with continual military conflict, terrorism, cyber-attacks, covert war, drone strikes, state-sanctioned rape, and millions of refugees. Isn't starving a people and destroying their topsoil, or imprisoning a nation's young black men, a form of violence?

Capitalism no more reduced violence than automobiles saved us from manure-filled cities. We may be less likely to be assaulted randomly in the street than we were in medieval times, but that doesn't mean humanity is less violent, or that the blind pursuit of continued economic growth and technological progress is consonant with the increase of human welfare—no matter how well such proclamations do on the business best-seller lists or speaking circuit. (Businesspeople don't want to pay to be told that they're making things worse.)

So with the blessings of much of the science industry and its collaborating futurists, corporations press on, accelerating civilization under the false premise that because things are looking better for the wealthiest beneficiaries, they must be better for everyone. Progress is good, they say. Any potential impediment to the frictionless ascent of technological and economic scale—such as the cost of labor, the limits of a particular market, the constraints of the planet, ethical misgivings, or human frailty—must be eliminated.

The models would all work if only there weren't people in the way. That's why capitalism's true believers are seeking someone or, better, something to do their bidding with greater intelligence and less empathy than humans.

ARTIFICIAL INTELLIGENCE

53.

In the future envisioned by Wall Street and Silicon Valley alike, humans are just another externality. There are too many of us, asking for salaries and healthcare and meaningful work. Each victory we win for human labor, such as an increase in the minimum wage, makes us that much more expensive to employ, and supports the calculus through which checkout workers are replaced by touchscreen kiosks

Where humans remain valuable, at least temporarily, is in training their replacements. Back in the era of outsourcing, domestic workers would cry foul when they were asked to train the lower-wage foreign workers who would shortly replace them. Today, workers are hardly aware of the way digital surveillance technologies are used to teach their jobs to algorithms.

This is what all the hoopla about "machine learning" is really about. The things we want our robots to do—like driving in traffic, translating languages, or collaborating with humans—are mind-bogglingly complex. We can't devise a set of explicit instructions that covers every possible situation. What computers lack in improvisational logic, they must make up for with massive computational power. So computer scientists feed the algorithms reams and reams of data, and let them recognize patterns and draw conclusions themselves.

They get this data by monitoring human workers doing their jobs. The ride-hailing app on cab drivers' phones also serves as a recording device, detailing the way they handle various road situations. The algorithms then parse data culled from thousands of drivers to write their own autonomous vehicle programs. Online task systems pay people pennies per task to do things that computers can't yet do, such as translate certain phrases, label the storefronts in photos, or identify abusive social media posts. The companies paying for the millions of human micro-tasks may not actually need any of the answers themselves. The answers are being fed directly into machine learning routines.

The humans' only real job is to make themselves obsolete.

54.

Losing one's job to a robot is no fun.

Without a new social compact through which to distribute the potential bounty of the digital age, competition with our machines is a losing proposition. Most jobs as we currently under-stand them are repetitive enough to be approached computation-ally. Even brain surgery is, in most respects, a mechanical task with a limited number of novel scenarios.

While we humans can eventually shift, en masse, to high-touch occupations like nursing, teaching, psychology, or the arts, the readiness of machines to replace human labor should force us to reevaluate the whole premise of having jobs in the first place.

Employment, as we currently understand it, emerged only in the late Middle Ages, when the peer-to-peer economy was dismantled. Monarchs gave out exclusive monopolies to their favorite companies, forcing everyone else to become employees of the chosen few. Instead of selling the value they created, former craftspeople and business owners now sold their *time*. Humans became resources.

The employment model has become so prevalent that our best organizers, representatives, and activists still tend to think of prosperity in terms of getting everyone "jobs," as if what everyone really wants is the opportunity to commodify their living hours. It's not that we need full employment in order to get everything done, grow enough food, or make enough stuff for everyone. In the United States, we already have surplus food and housing. The Department of Agriculture regularly burns crops in order to keep market prices high. Banks tear down houses that are in foreclosure lest they negatively impact the valuation of other houses and the mortgages they're supporting.

But we can't simply give the extra food to the hungry or the surplus houses to the homeless. Why? Because they don't have jobs! We punish them for not contributing, even though we don't actually *need* more contribution.

Jobs have reversed from the means to the ends, the ground to the figure. They are not a way to guarantee that needed work gets done, but a way of justifying one's share in the abundance. Instead of just giving people food and shelter, our governments lend easy money to banks in the hope that they will invest in corporations that will then build factories. The products they manufacture may be unnecessary plastic contraptions for which demand must be created with manipulative marketing and then space made in landfills, but at least they will create an excuse to employ some human hours.

If we truly are on the brink of a jobless future, we should be celebrating our efficiency and discussing alternative strategies for distributing our surplus, from a global welfare program to universal basic income. But we are nowhere close. While machines may get certain things done faster and more efficiently than humans, they externalize a host of other problems that most technologists pretend do not exist. Even today's robots and computers are built with rare earth metals and blood minerals; they use massive amounts of energy; and when they grow obsolete their components are buried in the ground as toxic waste. Plus, the bounty produced by modern technocapitalism is more than offset by its dependence on nonrenewable resources and human slavery.

By hiring more people rather than machines, paying them livable wages, and operating with *less* immediate efficiency, companies could minimize the destruction they leave in their wake. Hiring ten farmers or nurses may be more expensive in the short run than using one robotic tractor or caregiver, but it may just make life better and less costly for everyone over the long term.

In any case, the benefits of automation have been vastly overstated. Replacing human labor with robots is not a form of liberation, but a more effective and invisible way of externalizing the true costs of industry. The jobless future is less a reality to strive toward than the fantasy of technology investors for whom humans of all kinds are merely the impediment to infinite scalability.

55.

A future where we're all replaced by artificial intelligence may be further off than experts currently predict, but the readiness with which we accept the notion of our own obsolescence says a lot about how much we value ourselves. The long-term danger is not that we will lose our jobs to robots. We can contend with joblessness if it happens. The real threat is that we'll lose our humanity to the value system we embed in our robots, and that they in turn impose on us.

Computer scientists once dreamed of enhancing the human mind through technology, a field of research known as intelligence augmentation. But this pursuit has been largely surrendered to the goal of creating artificial intelligence—machines that can think for themselves. All we're really training them to do is manipulate our behavior and engineer our compliance. Figure has again become ground.

We shape our technologies at the moment of conception, but from that point forward they shape us. We humans designed the telephone, but from then on the telephone influenced how we communicated, conducted business, and conceived of the world. We also invented the automobile, but then rebuilt our cities around automotive travel and our geopolitics around fossil fuels.

While this axiom may be true for technologies from the pencil to the birth control pill, artificial intelligences add another twist: after we launch them, they not only shape us but they also begin to shape themselves. We give them an initial goal, then give them all the data they need to figure out how to accomplish it. From that point forward, we humans no longer fully understand how an AI may be processing information or modifying its tactics. The AI isn't conscious enough to tell us. It's just trying everything, and hanging onto what works.

Researchers have found, for example, that the algorithms running social media platforms tend to show people pictures of their ex-lovers having fun. No, users don't want to see such images. But, through trial and error, the algorithms have discovered that showing us pictures of our exes having fun increases our engagement. We are drawn to click on those pictures and see what our exes are up to, and we're more likely to do it if we're jealous that they've found a new partner. The algorithms don't know why this works, and they don't care. They're only trying to maximize whichever metric we've instructed them to pursue.

That's why the original commands we give them are so important. Whatever values we embed—efficiency, growth, security, compliance—will be the values AIs achieve, by whatever means happen to work. AIs will be using techniques that no one—not even they—understand. And they will be honing them to generate better results, and then using those results to iterate further.

We already employ AI systems to evaluate teacher performance, mortgage applications, and criminal records, and they make decisions just as racist and prejudicial as the humans whose decisions they were fed. But the criteria and processes they use are deemed too commercially sensitive to be revealed, so we cannot open the black box and analyze how to solve the bias. Those judged unfavorably by an algorithm have no means to appeal the decision or learn the reasoning behind their rejection. Many companies couldn't ascertain their own AI's criteria anyway.

As AIs pursue their programmed goals, they will learn to leverage human values as exploits. As they have already discovered, the more they can trigger our social instincts and tug on our heartstrings, the more likely we are to engage with them as if they were human. Would you disobey an AI that feels like your parent, or disconnect one that seems like your child?

Eerily echoing the rationale behind corporate personhood, some computer scientists are already arguing that AIs should be granted the rights of living beings rather than being treated as mere instruments or slaves. Our science fiction movies depict races of robots taking revenge on their human overlords—as if this problem is somehow more relevant than the unacknowledged legacy of slavery still driving racism in America, or the twenty-first-century slavery on which today's technological infrastructure depends.

We are moving into a world where we care less about how other people regard us than how AIs do.

56.

Algorithms do reflect the brilliance of the engineers who craft them, as well as the power of iterative processes to solve problems in novel ways. They can answer the specific questions we bring them, or even generate fascinating imitations of human creations, from songs to screenplays. But we are mistaken if we look to algorithms for direction. They are not guided by a core set of values so much as by a specific set of outcomes. They are utilitarian.

To a hammer, everything is a nail. To an AI, everything is a computational challenge.

We must not accept any technology as the default solution for our problems. When we do, we end up trying to optimize ourselves for our machines, instead of optimizing our machines for us. Whenever people or institutions fail, we assume they are simply lacking the appropriate algorithms or upgrades.

By starting with the assumption that our problems are fixable by technology, we end up emphasizing very particular strategies. We improve the metrics that a given technology can improve, but often ignore or leave behind the sorts of problems that the technology can't address. We move out of balance, because our money and effort go toward the things we can solve and the people who can pay for those solutions. We've got a greater part of humanity working on making our social media feeds more persuasive than we have on making clean water more accessible. We build our world around what our technologies can do.

Most technologies start out as mere tools. At first they exist to serve our needs, and don't directly contradict our worldview or our way of life. If anything, we use them to express our own, existing values. We built airplanes so humans could experience flight and travel great distances. We developed radio to extend our voices across space. Their primary impact on our world is to execute their original purpose.

However, as technologies become more a part of our world, we begin making more accommodations to their functioning. We learn to cross the street carefully so as not to be hit by automobiles, we clear-cut a forest to make way for electric cables, or we dedicate a room once devoted to conversation and family— the *living* room—to the television. The technology forces negotiations and compromises.

Without human intervention, technology becomes an accepted premise of our value system: the starting point from which everything else must be inferred. In a world of text, illiteracy is the same as stupidity, and the written law may as well be the word of God. In a world defined by computers, speed and efficiency become the primary values. Refusing a technological upgrade may as well be a rejection of the social norm, or a desire to remain sick, weak, and unrepentantly human.

57.

Human beings are not the problem. We are the solution.

To many of the developers and investors of Silicon Valley, however, humans are not to be emulated or celebrated, but transcended or—at the very least—reengineered. These technologists are so dominated by the values of the digital revolution that they see anything or anyone with different priorities as an impediment. This is a distinctly antihuman position, and it's driving the development philosophy of the most capitalized companies on the planet.

In their view, evolution is less the story of life than of data. Information has been striving for greater complexity since the beginning of time. Atoms became molecules, molecules became proteins, proteins became cells, organisms, and eventually humans. Each stage represented a leap in the ability to store and express information.

Now that we humans have developed computers and networks, we are supposed to accept the fact that we've made something capable of greater complexity than ourselves. Information's journey to higher levels of dimensionality must carry on beyond biology and humans to silicon and computers. And once that happens, once digital networks become the home for reality's most complex structures, then human beings will really only be needed insofar as we can keep the lights on for the machines. Once our digital progeny can care for themselves, we may as well exit the picture.

This is the true meaning of "the singularity": it's the moment when computers make humans obsolete. At that point, we humans will face a stark choice. Either we enhance ourselves with chips, nanotechnology, and genetic engineering to keep up with our digital superiors; or we upload our brains to the network.

If we go the enhancement route, we must accept that whatever it means to be human is itself a moving target. We must also believe that the companies providing us with these upgrades will be our trustworthy partners—that they wouldn't remotely modify equipment we've installed into ourselves, or change the terms of service, or engineer incompatibility with other companies' enhancements or planned obsolescence. Given the track record of today's tech companies, that's not a good bet. Plus, once we accept that every new technology has a set of values that goes along with it, we understand that we can't incorporate something into ourselves without installing its affordances as well. In the current environment, that means implanting extractive, growth-based capitalism into our bloodstreams and nervous systems.

If we go with uploading, we'd have to bring ourselves to believe that our consciousness somehow survives the migration from our bodies to the network. Life extension of this sort is a tempting proposition: simply create a computer as capable of complexity as the brain, and then transfer our consciousness—if we can identify it—to its new silicon home. Eventually, the computer hosting our awareness will fit inside a robot, and that robot can even look like a person if we want to walk around that way in our new, eternal life. It may be a long shot, but it's a chance to keep going.

Others are hoping that even if our consciousness does die with our body, technologists will figure out how to copy who we are and how we think into an AI. After that, our digital clone could develop an awareness of its own. It's not as good as living on, perhaps, but at least there's an instance of "you" or "me" somewhere out there. If only there were any evidence at all that consciousness is an emergent phenomenon, or that it is replicable in a computer simulation. The only way to bring oneself to that sort of conclusion is to presume that our reality is itself a computer simulation—also a highly popular worldview in Silicon Valley.

Whether we upload our brains to silicon or simply replace our brains with digital enhancements one synapse at a time, how do we know if the resulting beings are still alive and aware? The famous "Turing test" for computer consciousness determines only whether a computer can *convince* us that it's human. This doesn't mean that it's actually human or conscious.

The day that computers pass the Turing test may have less to do with how smart computers have gotten than with how bad we humans have gotten at telling the difference between them and us.

58.

Artificial intelligences are not alive.

They do not evolve. They may iterate and optimize, but that is not evolution. Evolution is random mutation in a particular environment. Machine learning, by contrast, is directed toward particular, preprogrammed ends. It may be complicated, but—unlike evolution, the weather, the oceans, or nature—it's not complex. Complicated systems, such as a city's many traffic lights, direct activity from the top down. By contrast, complex systems such as a traffic circle establish flows spontaneously through the interaction of their many participants. Machines have lots of complicated parts and processes, but no higher-order, lifelike complexity is emerging from them.

No matter how many neurological-sounding terms are coined for what computers do, they are not on the path to consciousness. Those fantasizing about computer life like to throw around terms like "fuzzy logic," as if these programming techniques are bringing machines closer to exhibiting human intuition. Fuzzy logic is the ability of a computer program to consider values other than 1 or 0, and then express them as a 1 or a 0. That's all. Fuzzy logic isn't fuzzy in the manner of true uncertainty. It merely reduces the roughness and complexity of reality into a simple binary that a computer can work with.

Similarly, neural nets are not like human brains. They're simply layers of nodes that learn to do things by being fed hundreds or thousands of examples. Instead of telling the computer what a cat looks like, we just feed it hundreds of pictures until it can determine the common, distinguishing features. Human brains are capable of generalizing a category like "cat" after seeing just one example. How? We're not exactly sure.

Our inability to say exactly what it means to be a thinking, autonomous human being should not be considered a liability. The human mind is not computational, any more than reality itself is just information. Intelligence is a fabulous capability of the brain, and reality stores immense quantities of data—but neither of these exists at all without human consciousness to render them. We must not reduce human awareness to raw processing power. That's akin to reducing the human body to weightlifting. Our calculation speeds can't compete with those of a supercomputer, and we will never lift as much as a crane. But humans are worth more than our utility. Enhancing one job-friendly metric with technological intervention or replacement just leaves other, likely more important values behind. The most important of those is consciousness itself.

As best we know, consciousness is based on totally noncomputable quantum states in the tiniest structures of the brain, called microtubules. There are so many billions of these microtubules, and then so many active, vibrating sites on each one, that a machine harnessing every computer chip ever made would wither under the complexity of one human brain.

The only people who behave as though consciousness were simple enough to be replicated by machine are computer developers. Real neuroscientists remain delightfully flummoxed by the improbability of self-awareness emanating from a lump of neurons. It's confounding and paradoxical.

That doesn't mean we will soon be able to dismiss consciousness. It's not some illusion perpetrated by DNA on the human brain in order to compel a survival instinct in its hosts. We do not live in a simulation; our awareness is real. When pressed, even physicists accept that consciousness has a better claim to existence than objective reality. Quantum theory holds that objective reality may not exist until we observe it. In other words, the universe is a bunch of possibilities until someone's consciousness arrives on the scene and sees it a certain way. Then it condenses into what we call reality.

The search for the seed of consciousness is a bit like the quest for the smallest cosmological particle. It's more an artifact of our mechanistic science than a reflection of how the mind works. Whenever we discover one ultimate determinant—such as the gene—we also discover that its expression is determined by something else. No sooner do we find the germ we deem responsible for a disease than we also find the environmental factors allowing it to flourish, or the immune deficiency, or the transition from helpful bacteria to invading pathogen. The only way to solve consciousness is through first-hand experience and reverence for the world in which we live, and the other people with whom we share it.

In this sense, we know consciousness exists because we know what it feels like. Just like an animal or a computer, we can see a cup of coffee sitting on the kitchen table. But we humans also know *what it is like* to see a cup of coffee on the table. The choice to look at that cup, to pay attention, is unique to consciousness. Computers can't do that. They have to see everything in range. They have no attention. No focus. No real direction.

And to know what it is like to look at a cup of coffee, to be able to consider that construction of mind and self: only humans can do that. That's because we're alive, and computers are not.

FROM PARADOX TO AWE

59.

You know that moment when a dog sees something it doesn't quite understand? When it tilts its head a little bit to one side, as if viewing the perplexing phenomenon from another angle will help? That state of confusion, that *huh?*, may be a problem for the dog, but it's awfully cute to us. That's because for humans, a state of momentary confusion offers not just frustration but an opening.

Team Human has the ability to tolerate and even embrace ambiguity. The stuff that makes our thinking and behavior messy, confusing, or anomalous is both our greatest strength and our greatest defense against the deadening certainty of machine logic.

Yes, we're living in a digital age, where definitive answers are ready at the click. Every question seems to be one web search away. But we are mistaken to emulate the certainty of our computers. They are definitive because they have to be. Their job is to resolve questions, turn inputs into outputs, choose between 1 or 0. Even at extraordinary resolutions, the computer must decide if a pixel is here or there, if a color is this blue or that blue, if a note is this frequency or that one. There is no in-between state. No ambiguity is permitted.

But it's precisely this ambiguity—and the ability to embrace it—that characterizes the collectively felt human experience. Does God exist? Do we have an innate purpose? Is love real? These are not simple yes-or-no questions. They're yes-*and*-no ones: Mobius strips or Zen koans that can only be engaged from multiple perspectives and sensibilities. We have two brain hemi-

spheres, after all. It takes both to create the multidimensional conceptual picture we think of as reality.

Besides, the brain doesn't capture and store information like a computer does. It's not a hard drive. There's no one-to-one correspondence between things we've experienced and data points in the brain. Perception is not receptive, but active. That's why we can have experiences and memories of things that didn't "really" happen. Our eyes take in 2D fragments and the brain renders them as 3D images. Furthermore, we take abstract concepts and assemble them into a perceived thing or situation. We don't see "fire truck" so much as gather related details and then manufacture a fire truck. And if we're focusing on the fire truck, we may not even notice the gorilla driving it.

Our ability to be conscious—to have that sense of what-is-it-like-to-see-something—depends on our awareness of our participation in perception. We feel ourselves putting it all together. And it's the open-ended aspects of our experience that keep us conscious of our participation in interpreting them. Those confusing moments provide us with opportunities to experience our complicity in reality creation.

It's also what allows us to do all those things that computers have been unable to learn: how to contend with paradox, engage with irony, or even interpret a joke. Doing any of this depends on what neuroscientists call relevance theory. We don't think and communicate in whole pieces, but infer things based on context. We receive fragments of information from one another and then use what we know about the world to re-create the whole message ourselves. It's how a joke arrives in your head: some assembly is required. That moment of "getting it"—putting it together oneself—is the pleasure of active reception. Ha! and Aha! are very close relatives.

Computers can't do this. They can recognize if a social

media message is sarcastic or not—yes or no—but they can't appreciate the dynamic contrast between word and meaning. Computers work closer to the way primitive, reptile brains do. They train on the foreground, fast-moving objects, and surface perceptions. *There's a fly; eat it.* The human brain, with its additional lobes, can also reflect on the greater spatial, temporal, and logical contexts of any particular event. *How did that fly get in the room if the windows have been closed?*

Human beings can relate the figure to the ground. We can hold onto both, and experience the potential difference or voltage between them. *The fly doesn't belong here.* Like the rack focus in a movie scene, we can compare and contrast the object to its context. We can ponder the relationship of the part to the whole, the individual to the collective, and the human to the team.

60.

Art, at its best, mines the paradoxes that make humans human. It celebrates our ability to embrace ambiguity, and to experience this sustained, unresolved state as pleasurable, or at least significant.

Commercial entertainment, by contrast, has the opposite purpose. The word *entertain*—from the Latin for "to hold within"—literally means "maintain," or "continue in a certain condition." Its goal is to validate the status quo values by which we already live, reinforce consumerism, and—most of all—reassure us that there is certainty in this world. Not only do we find out whodunnit, but we get to experience a story in which there are definitive answers to big questions, villains to blame when things go wrong, and a method for administering justice. These plots depict a character we like (usually a young man), put him in danger, raise the stakes until we can't take it anymore, and then give him the solution he needs to vanquish his enemy and win the day, at which point we can all breathe a sigh of relief. It's the stereotypically male arc of arousal: crisis, climax, and sleep.

This arc of tension and release, or complication and cure, has dominated our culture and defined not only our entertainment but our businesses, religions, and life journeys. Entrepreneurs don't want to create a company that succeeds by sustaining itself; they want a "home run" so they can sell it to someone else. Worshippers turn to religion less to explore their relationship to creation or ethics than to guarantee their own salvation or claim to virtue. We play life as a board game with winners and losers, where we "end up" in a certain career, marriage, and socioeconomic class.

We have been trained to expect an answer to every question, and an ending to every beginning. We seek closure and resolution, growing impatient or even despondent when no easy answer is in sight. This fuels capitalism and consumerism, which depend on people believing that they are just one stock market win or product purchase away from fulfillment. It's great for motivating a nation to, say, put a man on the moon before the end of the decade, or go to war against some other nation.

But it doesn't serve us as we attempt to contend with long-term, chronic problems. There's no easy fix for climate change, the refugee crisis, or terrorism. How do we even know when we're done? There's no flag to plant, no terms of surrender. Motivating a society to address open-ended challenges requires a more open-ended approach—one that depends less on our drive toward climax than on our capacity for unresolved situations. Like life.

It requires living humans.

61.

Prohuman art and culture question the value of pat narratives. They produce open-ended stories, without clear victors or well-defined conflicts. Everyone is right; everyone is wrong. The works don't answer questions; they raise them.

These are the "problem plays" of Shakespeare, which defy easy plot analysis, as characters take apparently unmotivated actions. They're the abstract paintings of Kandinsky or Delaunay, which maintain distance from real-world visual references. These images may represent figures, but only *sort of*. The observing human mind is the real subject of the work, as it tries and fails to identify objects that correspond perfectly with the images. And this process itself mirrors the way the human brain identifies things in the "real" world by perceiving and assembling fragmented details. Instead of giving us clear representation—this is an apple!—art stretches out the process of seeing and identifying, so that we can revel in the strange phenomenon of human perception.

We experience the same sorts of challenges watching the movies and television of David Lynch, who is likely to leave the camera on a character just sweeping the floor or smoking a cigarette for five minutes or more. Lynch is training us to let go of conventional story expectations so that we can learn to watch for something else—for the behavior of the humans in the scenes, for the activity that emerges out of boredom, and for the relationship of the characters to their world. Lynch is intentionally denying his audience the engagement that comes with tension and release, or even just plot. The various conspiracies in his stories don't even add up. That's because we're not supposed to be looking there.

As novelist Zadie Smith puts it, the writer's job is not to "tell us how somebody felt about something, it is to tell us how the world works." Such art no longer focuses on the protagonist and his heroic journey, but on the relationship of the figures to the ground. In doing so, it activates and affirms the uniquely human ability to experience context and make meaning.

Of course, the work of filmmakers, artists, and novelists creating in this way is emphatically countercultural—if for no other reason than that it questions traditional narratives and heroic, individualistic values. Any art that asks its viewers to slow down or, worse, pause and reflect is hurting a market that depends on automatic and accelerating behaviors. Film schools don't teach anti-plot, studios don't produce it (knowingly), and audiences don't commonly reward it. That's often used as an argument for its inferiority and irrelevance. If this stuff were really more human and deeply rewarding on some level, shouldn't it do better at the box office?

Commercial work with a central figure, rising tension, and a satisfying resolution succeeds because it plays to our fears of uncertainty, boredom, and ambiguity—fears generated by the market values driving our society in the first place. Moreover, we live in a world where uncertainty is equated with anxiety instead of with life. We ache for closure. That's why people today are more likely to buy tickets for an unambiguously conclusive blockbuster than for an anticlimactic, thought-provoking art film. It's because we've been trained to fear and reject the possibility that reality is a participatory activity, open to our intervention.

62.

The rise of digital media and video games has encouraged the makers of commercial entertainment to mimic some of the qualities of postnarrative work, but without actually subjecting their audiences to any real ambiguity.

Movies and prestige television, for example, play with the timeline as a way of introducing some temporary confusion into their stories. At first, we aren't told that we're watching a sequence out of order, or in multiple timelines. It's just puzzling. Fans of ongoing series go online to read recaps and test theories with one another about what is "really" going on. But by the end of the series, we find out the solution. There is a valid timeline within an indisputable reality; we just had to put it together. Once we assemble the puzzle pieces, the show is truly over.

In a nod to the subscription model of consumption—where we lease cars or pay monthly to a music service—the extended narratives of prestige TV series spread out their climaxes over several years rather than building to a single, motion picture explosion at the end. But this means energizing the audience and online fan base with puzzles and "spoilers." Every few weeks, some previously ambiguous element of the story is resolved: the protagonist and antagonist are two parts of a single character's split personality, the robots' experiences took place a decade ago, those crew members are really androids, and so on.

Spoilers, as their name implies, must be protected lest they spoil the whole experience for someone else. They're like land mines of intellectual property that are useless once detonated. We are obligated to keep the secret and maintain the value of the "intellectual property" for others. The superfan of commercial entertainment gets rewarded for going to all the associated websites and fan forums, and reading all the official novels. Superfans know all the answers because they have purchased all the products in the franchise. Like one of those card games where you keep buying new, expensive packs in order to assemble a powerful team of monsters, all it takes to master a TV show is work and money.

Once all the spoilers have been unpacked, the superfan can rewatch earlier episodes with the knowledge of what was "really" going on the whole time. No more damned ambiguity. The viewer gets to experience the story again, but with total knowledge and total control—as if omniscience were the desired state of mind, rather than a total negation of what makes humans conscious in the first place.

A show's "loose ends" are its flaws. They prevent the superfan from maintaining a coherent theory of everything. They are not thought of as delightful trailheads to new mysteries, but as plot holes, continuity errors, or oversights by the creators. In commercial entertainment, where the purpose is always to give the audience their money's worth, submission to the storyteller must be rewarded with absolute resolution. This same urge is driving such entertainment to ever higher frame rates and pixel counts—as if seeing the picture clearer and bigger is always better. We don't make sense of it; the sense is made for us. That's what we're paying for.

Loose ends threaten to unravel not only the fictions upholding an obsolete Hollywood format, but also the ones upholding an obsolete social order: an aspirational culture in which product purchases, job advancement, trophy spouses, and the accumulation of capital are the only prizes that matter.

Loose ends distinguish art from commerce. The best, most humanizing art doesn't depend on spoilers. What is the "spoiler" in a painting by Picasso or a novel by James Joyce? The impact of a classically structured art film like *Citizen Kane* isn't compromised even if we do know the surprise ending. These masterpieces don't reward us with answers, but with new sorts of questions. Any answers are constructed by the audience, provisionally and collaboratively, through their active interpretation of the work.

Art makes us think in novel ways, leading us to consider new approaches and possibilities. It induces states of mind that are often strange and uncomfortable. Rather than putting us to sleep, art wakes us up and invites us to experience something about being human that is in danger of being forgotten. The missing ingredient can't be directly stated, immediately observed, or processed by algorithm, but it's there—in the moment before it is named or depicted or resolved.

It's alive, it's paradoxical, and it's the exclusive province of Team Human

63.

While humans are drawn to and empowered by paradox, our market-driven technologies and entertainment appear to be fixed on creating perfectly seamless simulations.

We can pinpoint the year movies or video games were released based on the quality of their graphics: the year they figured out steam, the year they learned to reflect light, or the year they made fur ripple in the wind. Robot progress is similarly measured by the milestones of speech, grasping objects, gazing into our eyes, or wearing artificial flesh. Each improvement reaches toward the ultimate simulation: a movie, virtual reality experience, or robot with such high fidelity that it will be indistinguishable from real life.

It's a quest that will, thankfully, never be achieved. The better digital simulations get, the better we humans get at distinguishing between them and the real world. We are in a race against the tech companies to develop our perceptual apparatus faster than they can develop their simulations.

The hardest thing for animators and roboticists to simulate is a living human being. When an artificial figure gets too close to reality—not so close as to fool us completely, yet close enough that we can't tell quite what's wrong—that's when we fall into a state of unease known as the "uncanny valley." Roboticists noticed the effect in the early 1970s, but moviemakers didn't encounter the issue until the late 1980s, when a short film of a computer-animated human baby induced discomfort and rage in test audiences. That's why filmmakers choose to make so many digitally animated movies about toys, robots, and cars. These objects are easier to render convincingly because they don't trigger the same mental qualms.

We experience vertigo in the uncanny valley because we've spent hundreds of thousands of years fine-tuning our nervous systems to read and respond to the subtlest cues in real faces. We perceive when someone's eyes squint into a smile, or how their face flushes from the cheeks to the forehead, and we also—at least subconsciously—perceive the absence of these organic barometers. Simulations make us feel like we're engaged with the non-living, and that's creepy.

We confront this same sense of inauthenticity out in the real world, too. It's the feeling we get when driving past fake pastoral estates in the suburbs, complete with colonial pillars and horse tie rings on the gates. Or the strange verisimilitude of Las Vegas's skylines and Disney World's Main Street. It's also the feeling of trying to connect with a salesperson who sticks too close to their script.

In our consumer culture, we are encouraged to assume roles that aren't truly authentic to who we are. In a way, this culture is its own kind of simulation, one that requires us to make more and more purchases to maintain the integrity of the illusion. We're not doing this for fun, like trying on a costume, but for keeps, as supposedly self-realized lifestyle choices. Instead of communicating to one another through our bodies, expressions, or words, we do it through our purchases, the facades on our homes, or the numbers in our bank accounts. These products and social markers amount to pre-virtual avatars, better suited to game worlds than real life.

Most of all, the uncanny valley is the sense of alienation we can get from ourselves. What character have we decided to play in our lives? That experience of having been cast in the wrong role, or in the wrong play entirely, is our highly evolved BS detector trying to warn us that something isn't right — that there's a gap between reality and the illusion we are supporting. This is a setup, our deeper sensibilities are telling us. Don't believe. It may be a trap. And although we're not Neanderthals being falsely welcomed into the enemy camp before getting clobbered, we are nonetheless the objects of an elaborate ruse—one that evolution couldn't anticipate.

Our uneasiness with simulations—whether they're virtual reality, shopping malls, or social roles—is not something to be ignored, repressed, or medicated, but rather felt and expressed. These situations feel unreal and uncomfortable for good reasons. The importance of distinguishing between human values and false idols is at the heart of most religions, and is the starting place for social justice.

The uncanny valley is our friend.

64.

The easiest way to break free of simulation is to recognize the charade and stop following the rules of the game.

No, cheating doesn't count. Illegal insider trades and performance-enhancing drugs simply prove how far people are willing to go to win. If anything, cheating reinforces the stakes and reality of the game.

Transcending the game altogether means becoming a spoilsport—someone who refuses to acknowledge the playing field, the rules of engagement, or the value of winning. (Why win, anyway, if it's only going to end the game?) In certain non-Western cultures, the spoilsport is the shaman, who lives apart from the tribe in order to see the larger patterns and connections. In a world where a person's success is measured by career achievements, the spoilsport is the one willing to sacrifice commercial reward for social good. In a middle school where social media likes are the metric of popularity, the spoilsport is the kid who deletes the app or chooses not to own a phone at all. The spoilsport takes actions that make no sense within the logic of the game.

Such anomalous behavior challenges convention, breaks the conspiracy of conformity, and stumps the algorithms. AIs and other enforcers of social control can't follow what they can't categorize. Weirdness is power, dissolving false binaries and celebrating the full spectrum of possibility. Eccentricity opens the gray area where mutations develop and innovations are born.

We can assert our uniquely human sides through things like humor and pranks, music and magic—none of which can be appreciated or even understood by machines or markets. Comedy demands that we identify with someone else and recognize our plight in theirs. Music communicates through aesthetics, while art challenges our sense of self and our relationship to the outer world. Stage magic confounds our logical sensibilities, contrasting the way things appear with the way we know they should be, while spiritual magick explores the seemingly impossible relationship between our will and the workings of the universe.

Weirdness crashes boundaries, forcing us to see our complicity in reality creation: we break free of the imposed program and experiment with alternatives. None of our models are authoritative, but that's the whole point. We each live within the confines of our own reality tunnels, seeing very limited pictures of the real world. That's what we have in common. The best way to resolve the image is to bring more people and perspectives into play.

That's why, most of all, being a spoilsport is a social signal. It is a way of calling out to the others who have unplugged from their programming and are seeking to reclaim their humanity.

The cues help announce what distinguishes us, but beware: unconventional behaviors are quickly identified, copied, and then sold back to us as commodified identities. That's why being truly anomalous has to mean more than the adoption of a particular style or intersectional label. It's about finding people with whom to connect more deeply and recognizing that the cues we use to identify one another are just means to that greater end.

65.

After delivering a lecture at Berkeley, a 1960s' counterculture psychologist took questions from the audience. A young woman stood up to explain that she understood the deep connection between people and our collective responsibility for the world, but she didn't know what to do next. The psychologist answered, "Find the others."

Find the others. Restore the social connections that make us fully functioning humans, and oppose all conventions, institutions, technologies, and mindsets that keep us apart. Challenging the overt methods of separation is straightforward: reject the hate speech of racists, the zero-sum economics of oppression, and the warmongering of both tyrants and neoliberal hawks. Our internalized obstacles to connection, however, are more embedded and pernicious. And they all tend to have something to do with shame.

For instance, we are trained from an early age not to talk with other people about money. Our salaries and savings are considered to be as private as our medical histories. Why? The habit finds its roots in the ascent of former peasants. When the aristocracy realized they could no longer keep ahead of the rising middle class, they sought nonmonetary methods of indicating their status, such as nobility of birth. Unable to keep up with bourgeois styles of dress or home decor, aristocrats pushed for less ornate aesthetics. In a reversal of centuries of ostentatious living, it became classy to hide one's wealth, rather than to display it.

Today, it's still considered rude to ask someone how much money they make. In certain situations we're embarrassed if we make too little, and in others we're ashamed if we make too much. But the whole social convention of hiding one's wealth or lack of it has less to do with protecting one another's feelings than protecting the controlling power of our superiors.

The boss gives you a salary increase—so long as you don't tell anyone else about it. For if you do, everyone else will be asking for the same thing. If you maintain the secret, you're in cahoots with management, submitting to the same dynamic as an abused child who is paid in candy to keep quiet. The bribe is a bond based in shame, and the bond is broken only when the victim finds others in whom to confide—often people who have experienced the same abuse. The real power comes when they're ready to say it out loud, as a movement of people opposing such abuse.

Likewise, the power of unions isn't just collective bargaining, but the collective sensibility that unionizing engenders.

DOUGLAS RUSHKOFF

The crosstalk between workers breaks management's efforts to make them compete with one another over scraps. That's why taxi apps and internet errand platforms don't have features that allow the workers to converse with one another about their experiences. Crosstalk breeds solidarity, and solidarity breeds discontent.

Religions, cults, governments, and social media platforms all use the same tactics to control their members: learn an individual's secrets, sexual proclivities, or identity issues, and then threaten to use this information against them. It's movie stars' fear of being outed that keeps them beholden to the cults they've confessed to long after they would have otherwise moved on. Some cults use lie detectors to drill down into their targets' most personal, shameful secrets. But these technologies are just updated versions of the confessionals once used by churches to blackmail their wealthy parishioners, or to shame the poor ones into exploitative compliance.

The happy explosion of new genders, racial identities, and disability intersections flies in the face of social programming designed to stigmatize difference and disadvantage those who can be labeled outsiders.

Shame does have a social function. Shaming those who deviate from the norm helps galvanize unity among the group and enforce adherence to the rules. Frat houses shame new recruits into macho antics, just as pious hypocrites shame their followers into obedience. In more prosocial hands, the same shaming tactics can be used by schools to stigmatize bullying, or by environmentalists to punish polluters. The problem is that people and institutions behaving destructively are not so vulnerable to shame. Bullies are proud of their conquests, and corporations experience no emotions.

Social stigma only truly hurts humans who are being human.

It is a counterproductive way of bonding people. Human teams should be based on common hopes, needs, strengths, and vulnerabilities. We don't get that by enforcing shame, but by embracing openness.

The internet, with its sometimes forced transparency, creates possibilities for the dissolution of shame, and for new bonds of solidarity across formerly impenetrable boundaries. It's no coincidence that a digital culture with imposed surveillance and inescapable exposure has also brought us gay marriage and cannabis reform.

The things people do become normal when they can't be shamed into silence about doing them.

66.

Once we dispense with shame, we are liberated to experience the full, sacred, unlikely wackiness of being human. We are confident enough leave the safety of the private computer simulation and jump into the wet chaos of social intimacy. Instead of marveling at the granularity of a VR world or the realism of a robot's facial expression, we open our senses to the taste of the breeze or the touch of a lover.

We exchange the vertigo of the uncanny valley for the exhilaration of awe.

The state of awe may be the pinnacle of human experience. It's what lies beyond the paradox. If humans' unique job in nature is to be conscious, what more human thing can we do than blow our observing minds? Beholding the panoramic view from a mountaintop, witnessing the birth of a child, staring into a starry sky, or standing with thousands of others in march or celebration, all dissolve the sense of self as separate and distinct. We experience ourselves as both the observing eye and the whole of which we are a part. It's an impossible concept, yet an undeniable experience of power and passivity, awareness and acceptance.

Psychologists tell us that the experience of awe can counteract self-focus, stress, apathy, and detachment. Awe helps people act with an increased sense of meaning and purpose, turning our attention away from the self and toward our collective self-interest. Awe even regulates the cytokine response and reduces inflammation. New experiments have revealed that after just a few moments of awe, people behave with increased altruism, cooperation, and self-sacrifice. The evidence suggests that awe makes people feel like part of something larger than themselves, which in turn makes them less competitive and more attuned to the needs of those around them.

Unfortunately, opportunities to experience awe in modern society are becoming more scarce. People spend less time camping or in nature, the night sky is polluted with light, and participation in the arts and culture is down. Art and outdoors classes in public schools have been jettisoned in favor of those that prepare students for the standardized tests on which schools are judged. There are no easy metrics for awe.

Like any extreme state of being, awe can also be exploited. Movies use special effects and giant spectacle scenes to leverage awe at specific moments in a story arc. Dictators hold huge rallies to exhilarate their followers while avoiding any reasoned debate. Even shopping malls attempt to generate a sense of awe with high ceilings and giant fountains. For a moment, awe overwhelms the senses and wipes the mind clean, making it more open to new input. This helps a person take in new information, but also makes them more vulnerable to manipulation. And once burned by someone manipulating awe, we are twice shy to open ourselves to it again. We become jaded and cynical as a defense against being wonderstruck.

Still, just because awe can be abused doesn't mean we should give up on its humanizing potential. There is a difference between real awe and manipulated excitement—between staring out onto the expanse of the Grand Canyon and standing in a sea of true believers at a nationalist rally. The manufactured brand of awe doesn't unify; it divides us into individual consumers or followers. We become fragmented, each imagining our own relationship to Dear Leader. True awe comes with no agenda. It's not directed toward some end or plan or person; there's no time limit or foe to vanquish. There is no "other."

True awe is timeless, limitless, and without division. It suggests there is a unifying whole to which we all belong—if only we could hold onto that awareness.

SPIRITUALITY AND ETHICS

67.

Modernity is not particularly conducive to awe. We are painfully disconnected from the larger cycles of day, night, moon, and season, making it harder for us to witness or identify with the inspiring renewal all around us. Spirituality has become less of a state of being than yet another goal to attain in the future.

Unlike today, the vast majority of humankind's experience was spent understanding time as circular. Only recently did we adopt a more historical approach to time, and a correspondingly more aggressive way of manifesting our spiritual destiny. That's the main difference between the spiritual systems that humans lived with over many millennia and the infant religions that fueled colonialism in the last dozen or so centuries.

In a cyclical understanding of time, the consequences of one's actions can never be externalized or avoided. Everyone reincarnates, so if you do something bad to another person, you'll have to meet them again. If you spoil the natural world, you will be reborn into it yourself. Time and history are nonexistent, and the individual is living in the constant present. As a result, everything and everyone is interdependent and emanating from the same, shared source of life.

The invention of writing gave people the ability to record the past and make promises into the future. Historical time was born, which marked the end of the spirituality of an eternal present, and the beginning of linear religion and monotheism. Before the notion of a past and a future, it was difficult to explain how a single, all-powerful god could exist if there was still so much wrong with Creation. With the addition of history, the imperfect world could be justified as a work in progress. God was perfect, but his plan for the world was not yet complete. Someday in the future, the messianic age would come, when God's perfection would be manifest. Those who were faithful or remained on the good side of God's law would end up okay in the end. The Bible was both the chronicle of a people's emergence from slavery and a contract with God—a covenant—for their future prosperity if they followed his commandments.

And so the duality of "before and after" became a central premise of religion. Things were now moving in one direction. Instead of the wholeness and possibility of a timeless, interconnected universe, the world of scripture had a timeline and destiny—at least for the faithful. The future was a work in progress for people and their God. Tomorrow would be better than today. Reincarnation became obsolete: only one messiah needed to die for all to resurrect. When he comes back, the whole thing is over. This was now a story with a beginning, a middle, and an end.

There were some positive repercussions to this more linear understanding of time. It provoked an entirely new approach to ethics and progress. We humans could make the world a better place, and move toward greater justice. As recounted by biblical legend, once the ancient Israelites escaped from the stasis of slavery they were liberated into a story of their own making. Moses and his brother-in-law immediately wrote the laws by which this new ethical people would live. Religion went from a set of timeless rituals to a directed behavioral code.

On the other hand, our focus on the future enabled our intended ends to justify almost any means. Inhumane disasters like the Crusades as well as the progressive philosophies of Hegel and Marx all depended on a teleological view of our world. At their best, these approaches elevate our commitment to ethics and social justice. But they also tend to divorce us from the present. We become able to do violence now for some supposedly higher cause and future payoff.

We drive forward with our eyes on the horizon, ignoring the devastation we create in our wake. We clear forests permanently, and extract coal, oil, and water that can't be replenished. We treat the planet and people as resources to be used up and thrown away. We enslave human beings to build luxury technologies, and subject people in faraway places to pollution and poverty. Corporations dismiss these devastating side effects as externalities—the collateral damage of doing business, falling entirely on people and places unacknowledged on their spreadsheets.

A belief in reincarnation or karma would make it hard to engage in such inhumanity without some fear of repercussion. Nothing can be externalized because everything comes back around. With reincarnation expunged from religion, we don't have to worry about someday meeting the person we harm today. By maintaining our belief in a divine intervention, we are more at liberty to destroy the natural world and await rescue from above.

A circular understanding of time is incompatible with such extraordinary, singular moments as an apocalypse. Everything just is, and has always been. There's no such thing as progress—just seasons and cycles. In fact, a common tenet of many pre-Judaic religions is that human beings can't take a genuinely original action. Human activity is experienced, instead, as the endless repetition of archetypal gestures. Any action taken by a person, any object created, is significant only insofar as it resonates in a greater, transcendent reality. Actions take on meaning because they reenact the divine. Every time people make love, they are reenacting the union of divine archetypes. Every thing made or constructed is just an echo of the gods' creativity.

To those of us enmeshed in modernity, this may sound positively boring and devoid of purpose. There's no emphasis on progress. There's no originality, no authorship, no copyright, and no patents. No direction.

But it's also entirely more sustainable than a unidirectional flow of natural resources into waste products—externalities that are supposed to be ignored until the end. Such a process runs counter to the regenerative principles of nature and existence.

People used to believe in circles. They came to believe in lines.

68.

Polytheism allowed people to embody the divine and feel connected to the unfolding of the universe. Their belief in reincarnation and the cyclical nature of time guaranteed that everything would remain interdependent and accountable to everything else.

Judaism replaced an embodied experience of the divine with abstract monotheism. In monotheism you don't reenact the divine so much as worship God and follow his rules. Christianity replaced the circularity of reincarnation with the linearity of salvation. We have fallen from the paradise of timeless grace, live in sin, and pray for redemption.

There are pros and cons to this shift in belief. A civilization of laws, innovation, and a commitment to progress can still be balanced with reverence for nature, the cycles of life, and the divine moment. But once the progressive, linear expectations of the monotheist religions dovetailed with the expectations of capitalism, that balance was lost. It was replaced with an entirely triumphalist drive for growth, expansion, and extraction.

On encountering the destructiveness of European colonialists, Native Americans concluded that the invaders must have a disease. They called it *wettiko*: a delusional belief that cannibalizing the life force of others is a logical and morally upright way to live. The Native Americans believed that wettiko derived from people's inability to see themselves as enmeshed, interdependent parts of the natural environment. Once this disconnect has occurred, nature is no longer seen as something to be emulated but as something to be conquered. Women, natives, the moon, and the woods are all dark and evil, but can be subdued by man, his civilizing institutions, his weapons, and his machines. Might makes right, because might is itself an expression of the divine.

Wettikos aren't limited exclusively to Europeans. Clearly, the tendency goes at least as far back as sedentary living, the hoarding of grain, and the enslavement of workers. Wanton destruction has long been recognized as a kind of psychic malady. It's the disease from which Pharaoh of biblical legend was suffering—so much so that God was said to have "hardened his heart," disconnecting him from all empathy and connection with nature. Pharaoh saw other humans as pests to be exterminated, and used his superior technologies—from agriculture to chariots—to bend nature to his divine will.

Both Judaism and Christianity sought to inoculate themselves from the threat of wettiko. Their founding priests understood that disconnecting from nature and worshipping an abstract God was bound to make people feel less empathetic and connected. Judaism attempted to compensate for this by keeping God out of the picture—literally undepicted. The Israelites had just escaped the death cults of Egypt, and developed an open-source approach to religion that involved constant revision by its participants. Even the letters of sacred texts were written in a script meant to look as transparent as flame. Unlike the arks at which they had worshipped in Egypt, the Israelites' ark was to have no idol on the top. Rather, they venerated an explicitly empty space, protected by two cherubim. The removal of idols allows people to see the divine in one another instead. Law even dictates that people can read Torah only with a *minyan*, a group of ten peers, as if to guarantee that worship is social.

Christianity, likewise, sought to retrieve the insight that a religion is less important as a thing in itself than as a way of experiencing and expressing love to others. The new version of Judaism turned attention away from the written law, which had become an idol of its own, and again toward the heart. Christ of the Bible was attempting to prevent religion from becoming the figure instead of the ground.

But the crucifix became an emblem of divine conquest, first in the Crusades, and later—with the advent of capitalism and industrialism—for colonial empires to enact and spread wettiko as never before. And the law, originally developed as a way of articulating a spiritual code of ethics, became a tool for chartered monopolies to dominate the world, backed by royal gunships. While Europeans took their colonial victories as providential, Native Americans saw white men as suffering from a form of mental illness that leads its victims to consume far more than they need to survive, and results in an "icy heart" incapable of compassion.

Clearly, the wettiko virus prevailed, and the society that emerged from this aggressive extraction still uses the promise of a utopian future to justify its wanton exploitation of people and nature in the present.

69.

Many Westerners have come to understand the problems inherent in a society obsessed with growth, and have struggled to assert a more timeless set of spiritual sensibilities. But, almost invariably, such efforts get mired in our ingrained notions of personal growth, progress and optimism.

Frank Baum, the author of *The Wizard of Oz*, embodied this dynamic. He was not only a devoted follower of the Russian spiritualist Madame Blavatsky but also the founder of the first magazine on window dressing and retail strategies for department stores. Dorothy's journey down the Yellow Brick Road combined the esoteric wisdom of his teacher with the can-do optimism of early twentieth-century American consumerism. The gifts Dorothy and her companions finally receive from the Wizard merely activate the potentials they had with them all along. All they really needed was a shift in consciousness, but good products and salesmanship didn't hurt. Similarly, Reverend Norman Vincent Peale's "positive thinking" derived from occult and transcendentalist roots, but caught on only when he framed it as a prosperity gospel. He taught the poor to use the power of prayer and optimism to attain the good life, and helped the wealthy justify their good fortune as an outward reward for their inner faith.

Vulnerable to the same ethos of personal prosperity, the counterculture movements of the 1960s and 1970s originally sought to undermine the religion of growth on which American society was based. Hippies were rejecting the consumerist, middle-class values of their parents, while scientists were taking LSD and seeing the Tao in physics. A new spirit of holism was emerging in the West, reflected in the lyrics of rock music, the spread of meditation and yoga centers, and the popularity of Buddhism and other Eastern religions. It appeared to herald a new age.

But all of these spiritual systems were being interpreted in the American context of consumerism. Herbs, seminars, and therapies were distributed through multilevel marketing schemes and advertised as turnkey solutions to all of life's woes. The resulting New Age movement stressed individual enlightenment over communal health. It was the same old personal salvation wine, only in California chardonnay bottles. The social justice agenda of the antiwar and civil rights movements was repackaged as the stridently individualistic self-help movement. They adopted psychologist Abraham Maslow's "hierarchy of needs" as their rubric, with "self-actualization" as the ultimate goal at the top of the pyramid. Never mind that Buddhists would have challenged the very premise of a self. The LSD trip, with its pronounced sense of journey, peak, and return, became the new allegory for individual salvation.

Wealthy seekers attended retreats at centers such as Esalen Institute, where they were taught by the likes of Maslow, Fritz Perls, Alan Watts, and other advocates of personal transformation. While there was certainly value in taking a break from workaday reality and confronting one's demons, the emphasis was on transcendence: if traditional religions taught us to worship God, in this new spirituality we would *be* as gods. It wasn't really a retrieval of ancient holism, timelessness, and divine reenactment at all so much as an assertion of good old linear, goal-based, ascension—practiced on the former sacred grounds of the Esselen Indians.

70.

The ultimate goal of personal transcendence was to leave the sinful, temporary body behind and float as a free, perfected consciousness. All that prior consumption was just fuel for the rocket, and the regrettable destruction something to leave behind with the rest of physical reality.

This wasn't a break from consumer capitalism, but the fulfillment of its ultimate purpose. And, by the 1970s, many other sectors and industries seemed to be coalescing around the same set of potentials. Computer scientists were pondering artificial intelligence; personal transformation gurus were helping people walk over burning coals; specially phased audio tapes helped people journey out of body. Mind over matter became the mantra. People sought to escape biological limits before middle age, not to mention death.

As the beating heart of this spiritual, technological, and cultural innovation, the Bay Area began to attract global attention from those looking to overturn or escape the established order. Wealthy progressives believed that they could apply the insights they had gained in their personal spiritual journeys to the world at large. They initiated projects to end hunger, cure cancer, communicate with animals, and contact aliens. Their most significant undertaking was to address the nuclear standoff between the United States and Soviet Union by holding a series of meetings between the two nations' most spiritually awakened and politically connected individuals.

The Soviet–American citizen diplomacy program brought together America's leading spiritual teachers, scientists, and psychologists with those of the Soviet Union. The Russian equivalent of New Age spirituality was represented by the practitioners of cosmism, a form of gnosticism that grew out of the Russian Orthodox tradition's emphasis on immortality. The cosmists were a big hit. Their pursuit of life extension technologies quickly overtook geopolitics as the primary goal of the conferences. Believing that not only could human beings transcend our mortality but that we could bring physicality with us—at least in some form—the cosmists convinced America's LSD-taking spiritualists that technology could give them a way of beating death.

The cosmists' original idea held that one could resurrect the dead by rearranging their atoms to the precise positions they were in while the person was alive. But the cosmists were working on other solutions, such as perfecting humans through intentional evolution, moving human consciousness into the bodies of robots, conquering death, colonizing space, or uploading ourselves to computers.

Such were the origins of today's transhumanist movement. These conferences were formative experiences for Silicon Valley's most influential executives, investors, professors, scientists, and technologists—some of whom founded the biggest digital companies in the world. This vision still motivates the development of artificial intelligence, private space exploration, robotics, data surveillance, and life extension.

Transhumanism exalts and preserves one particular expression of humanity, while leaving the rest of messy creation behind—or even exploiting it—in order to escape before the body dies or the world ends.

71.

The transhumanist movement is less a theory about the advancement of humanity than a simple evacuation plan. Techno-utopians like to think of themselves as orchestrating a complete break from civilization—a leap into outer space, cyberspace, machine consciousness, or artificial life. But their ideas just extend our same blind addiction to consumption, destruction, progress, and colonization. Cyber-wettiko.

European colonialists ignored the peoples and places they overran in their conquest of the planet in the belief that they were working toward some greater endpoint, some ordained destiny. If some people were exploited, or their ecosystems annihilated, well, you can't make an omelet without breaking a few eggs. Likewise, you can't grow consciousness on a computer chip without leaving a few people behind—or enslaving some laborers to procure the rare earth metals required.

The singularitans don't mind harming humans when necessary, because humanity as we know it will not really be participating in the brave new future. These bodies in which we live may as well be forests, oil fields, or any other natural resource that can be clear-cut, burned for fuel, or otherwise trashed in the name of our own immortality. One Bay Area startup harvests the blood of young people so that wealthy executives can transfuse new life into their aging bodies—at least, until they figure out how to upload their ever-so-valuable minds to the cloud.

Once computers and robots can do the thinking, humans won't even be necessary. It won't matter so much that we will have destroyed the natural environment, since neither robots nor uploaded humans will require it for sustenance. We will have evolved or engineered beyond the body—or developed an artificial intelligence that makes us humans irrelevant anyway. At that point, we will really only be required to service the machines—at least, until machines get better at doing that for themselves.

Then, say the chief scientists at the world's biggest internet companies, humanity may as well pass the torch to our evolutionary successors and get off the stage. Proponents of the singularity discount human objections as hubris, ego, or nostalgia. Humans are the problem; their eyes are part of that same evil, ambiguous natural world as women, forests, and uncivilized natives. Heck, humans *are* the natives, subject to unpredictable ebbs and flows of emotions and hormones and irrational needs.

Singularitans consider technology more trustworthy than

humans. Surveillance is not just a profit center for social media, but a way of safeguarding digital society from human resistance. Code enforces rules without bias (unless, of course, you happen to be the coder). It's a blockchain reality, where machines execute the letter of the law without any sense of the spirit. So much the better to accelerate development and reach the singularity before the clock runs out on the habitable biosphere.

Computers and artificial intelligences are pure intention, not clouded or tempered by human social priorities or moral misgivings. They are a direct, utilitarian extension of our apocalyptic urge to colonize the natural world.

We have finally found a way of inflicting wettiko on ourselves.

72.

Our addiction to expansion, growth, and transcendence derives from our hubris and need for control. We mistake colonizing a new region of the planet or dominating some aspect of nature for an expression of our creative power. We act as if we were gods, capable of creation and immune from the value systems that might otherwise restrain our will. And because this path is ultimately so unsatisfying, it is also addictive.

Like initiates in a twelve-step program, those of us suffering from wettiko must turn to a higher power if we want to stop our destructive behavior. It's difficult for many of us to believe in God, much less some divine wisdom or order to the universe. We may never accept the pre-historic sensibility that everything we do merely reenacts some archetypal gesture of a deity. But it's probably necessary that we at least accept that humans are best guided by some higher, universal ideals. These ideals may be preexisting laws of reality or principles of life with which we resonate. They could be morals discovered or invented by human beings through some innate sense of good. Whatever their origins, we need these ideals to guide our choices.

If we're not going to follow the commands of a king, a CEO, or an algorithm, then we need unifying values in order to work together as a team toward mutually beneficial goals. Even that idea—the notion that things should be mutually beneficial—is itself a higher-order value. It's an assumption about what's right, baked into not just our evolutionary history but also into the structure of a moral universe.

We do have a more profound sense of right and wrong than is supported by the functional logic of productivity or capitalism. Those are functional values, but they don't really inform an ethic. They don't animate us in any real way, or give us an alternative to the self-destructive path we're pursuing. They lead us to dominate nature, which ultimately includes subjugating the natural within ourselves. Our goals for human happiness and well-being become metrics within an industrial or economic system—over-rationalized and disconnected from the natural world.

We need a Reason for what we do: enduring values toward which we strive. These are not the reasons we do things—the practical, utilitarian purposes for our many activities—but the big Reason to be taking action at all. For example, the reasons for education are certainly important. Students gain skills, increase their cognitive abilities, and absorb facts. But the Reason for education? Learning, period. It is an ideal in itself.

What matters is that without Reasons, we are subject to an entirely utilitarian logic, which doesn't leave much room for humans or nature. It's a logic that tells us humans to just be *reasonable* and submit to compromise, rather than Reasoned and principled. It's the logic of might makes right, where utilitarian power outweighs greater good.

The ideals through which we combat such logic are not distanced or abstract. They're as close to the core of our being as Peace, Love, Unity, and Respect. They can't always be justified or rationalized, but that's not because they don't exist. We have simply lost our capacity to relate to them.

This innate, natural, effortless connection to ideals was surrendered to the market, to colonialism, to slavery, to extraction, and to technology, then justified with applied science, utilitarianism, and public relations. We reduced ideas to weaponized memes, and humankind to human resources. We got carried away with our utilitarian capabilities, and lost touch with the Reasons to exercise those capabilities in the first place. That's how we came to see ourselves as separate from nature, and capable of bending reality itself to our will no matter the cost to everyone and everything in our way.

It's time to rebalance our reasons with Reason, and occupy that strange, uniquely human place: both a humble part of nature, yet also conscious and capable of leaving the world better than we found it.

NATURAL SCIENCE

73.

It's tempting to declare war on the institutions and technologies that seek to remake our world in their own image. Radical environmentalists believe that the only way for nature to reassert itself is for human civilization to reduce its numbers and return to preindustrial conditions. Others believe it's too late, that we've already cast our lot with technological progress, genetic engineering, and global markets. In their view slowing down the engines of progress will merely prevent us from finding the solutions we need to fix our current crises.

Neither approach will work. We cannot dominate nature for much longer, but neither can we retreat from civilization. This cannot be a war between those who want to preserve nature and those pursuing progress. Team Human includes everybody.

If we respond to crisis in a polarized way, we surrender to the binary logic of the digital media environment. We become the thing we are resisting. Technology may have created a lot of problems, but it is not the enemy. Neither are the markets, the scientists, the robots, the algorithms, or the human appetite for progress. But we can't pursue them at the expense of more basic, organic, connected, emotional, social, and spiritual sensibilities, either. Instead, we must balance our human need to remain connected to nature with our corresponding desire to influence our own reality. It's not an either/or, but a both/and. It's not even a paradox.

We can be utterly in charge of the choice *not* to be utterly in charge. We can be fully human without being in complete control of our world. We may be nature's great observers and problem-solvers, but nature is not a problem to be solved. We must instead learn to work with nature, just as we must learn to work with the many institutions and technologies we have developed over the last millennium or so. We can't go back; we must go *through*.

This is the first lesson a whitewater rafter learns on encountering the rapids. As the raft enters the turbulence and begins to buck, the temptation is to resist the current by jamming the paddle straight down into the water and holding it as still as possible. Another temptation is to remove the paddle altogether and surrender to the current. Both strategies put the raft at the mercy of the river. The best response is to paddle harder and faster. Go *with* the current, making the necessary adjustments to avoid rocks and other obstacles. It's neither resistance nor passivity, but active participation: working in concert with what's happening to make it downriver in one piece.

We must learn to see the technologically accelerated social, political, and economic chaos ahead of us as an invitation for more willful participation. It's not a time to condemn the human activity that's brought us to this point, but to infuse this activity with more human and humane priorities. Otherwise, the technology—or any system of our invention, such as money, agriculture, or religion—ends up overwhelming us, seemingly taking on a life of its own, and forcing us to prioritize its needs and growth over our own.

We shouldn't get rid of smartphones, but program them to save our time instead of stealing it. We shouldn't close the stock markets, but retool them to distribute capital to businesses that need it instead of enslaving companies to the short-term whims of investors. We shouldn't destroy our cities, but work to make them more economically and environmentally sustainable. This will require more human ingenuity, not less.

Environmentalism sometimes makes us feel like we humans

are the problem. But we're not. Humans are not a cancer on the planet. At the same time, we can't ignore the fact that people are willful beings, capable of changing the natural environment to suit our whims and prone to dominating anything we deem threatening.

We have to apply that same sense of proactive problem-solving to the mortal threats of our own creation. The very same things we might do to prepare for a global catastrophe could also make us resilient enough to prevent one. Distributed energy production, fairer resource management, and the development of local cooperatives would both benefit survivors of a calamity and help reduce the stresses that could bring one on.

Now is not the time to abandon our can-do optimism, but to direct it toward priorities greater than world domination.

74.

There's no single answer.

Our current approaches to stewarding nature are vastly oversimplified. They may be complicated, but they're not complex. They have many moving parts, use lots of technology, and involve many different actors, but they don't acknowledge the interconnectedness of nature or the possibility of unseen factors. Nature is dynamic and in constant flux, so we must manage it with approaches that seek balance rather than the extraction of maximum yield.

The first hunter-gatherers to plant seeds couldn't have foreseen an era when the practice of agriculture would threaten the viability of the planet's topsoil. Nor should they be blamed. Agriculture, along with writing and cities, launched what we now think of as civilization. But agriculture also reduced the biodiversity of the human diet, making our food supply vulnerable to blight, pests, and vermin. Agriculture made us more sedentary and more susceptible to infection. Our skeletons got smaller and our livestock became more diseased. In many ways, hunter-gatherers enjoyed healthier, more sustainable living than the farmers who succeeded them.

As biblical legend reminds us, agriculture created the conditions for "feast and famine," or surpluses and deficiencies, which in turn offered opportunities for wealth and control. To people looking for power, agriculture provided a ready way of centralizing necessary resources—necessitating the inventions of both text and math.

Agriculture augured a new approach to the world: instead of gathering what the earth provides, farmers break the ground and grow what they want. Agriculture turns harvest from a gift of nature to an achievement of people. We have understood this irony for millennia. In biblical legend, Cain's "sacrifice" of the grain he grew and harvested was not accepted by God. Abel, a shepherd, humbly acknowledged that he was sacrificing an animal he did not create. Cain's cultivated crop, on the other hand, was rejected as hubris, for only God can create things.

But even with those mythical lessons in hand, we were unable to shake off agriculture's bias for monopoly. By the Middle Ages, when the last of Europe's common lands were enclosed by chartered monopolies, agriculture's worst liabilities were amplified. A society built on privatized farming became about control, extraction, and ownership, even at the expense of true efficiency, human health, and environmental sustainability.

Agriculture became the means to an end that had nothing to do with feeding people and everything to do with amassing power. Industrialized cotton farming in the American colonies became justification for the extremely lucrative slave trade. Industrialized agriculture today primarily serves the shareholders of chemical, pesticide, and bioengineering companies. Industrial farm advocates argue that organic practices don't scale because they are too intensive, but this is only true for the first year or two, while the soil destroyed by decades of chemical abuse is restored to health. Organic, biodiverse farming is not a luxury for the wealthy, but a path to survival for those who are currently starving in sub-Saharan Africa. We know now, beyond any doubt, that industrial agriculture gets less food out of the ground, with fewer nutrients, less efficiently, more expensively, and with greater environmental devastation than small and organic farming. This is no longer a debate.

Industrial agriculture succeeds by externalizing its true costs to others. It spawns high-cost diseases, both directly through contaminated food and livestock, and indirectly in the form of malnutrition, obesity, and diabetes. The fast food and corporate grocery industries, meanwhile, externalize the cost of transport to the public road system, and subjugation of supplier nations to the military—all while receiving anticompetitive subsidies from the government. Studies by the United Nations and the World Bank have concluded that genetic engineering has no positive role in meeting the world's food needs.

The problem with relying entirely on industrial approaches to the land, or anything, is that they oversimplify complex systems. They ignore the circulatory, regenerative properties of living organisms and communities, and treat everything in linear terms: inputs and outputs.

They might maximize one season's yield of crop, but at the expense of the soil matrix, nutrient levels, crop health, and future harvest yield. This then necessitates the use of more chemicals and genetic modifications, and the cycle continues.

By current estimates, the earth will run out of topsoil (the layer of earth in which plants can grow) within sixty years. That's great for markets based on scarcity, but terrible for a planet of humans who need to eat.

Agriculture is not just a zero-sum game of extracting value from the earth and monopolizing it. It means participating as both members and beneficiaries of a complex cycle of bounty.

75.

The planet's complex biosphere will survive us, one way or the other. Our own continuing participation, however, is in some doubt. Our aggressive industrial processes don't just threaten the diversity of other species; they threaten us, too. Increasing levels of carbon dioxide lead to sharp declines in cognitive ability. Global warming not only displaces huge populations, but the higher temperatures can lead to everything from the spread of disease to increased social unrest.

We are part of a complex system of feedback loops and interconnections, and must learn to approach our world with greater sophistication, empathy, and vision—not as short-sighted profiteers but as humans with legacies. The earth doesn't have to reject us as if we were an invading pathogen. Our consumption of resources need not deplete anything. In fact, we humans are capable of leaving a place more fertile than we found it.

One great model for how humans can participate willfully yet harmoniously in the stewardship of nature and resources is called permaculture. When the term was coined by a graduate student in 1978, it was meant to combine "agriculture" with "permanent." But it was expanded to mean "permanent culture," as a way of acknowledging that any sustainable approach

DOUGLAS RUSHKOFF

to food, construction, economics, and the environment had to bring our social reality into the mix.

Permaculture is a philosophy of working with, rather than against, nature. It means observing how plants and animals function together, rather than isolating one product or crop to extract. It involves recognizing the bigger, subtle cycles of the seasons and the moon, and treating them as more than superstition. It requires us to recognize earth as more than just dirt, but as soil: a highly complex network of fungi and micro-organisms through which plants communicate and nourish one another. Permaculture farmers treat soil as a living system, rather than "turning" it with machines and pulverizing it into dirt. They rotate crops in ways that replenish nutrients, make topsoil deeper, prevent water runoff, and increase speciation. They leave the landscape more alive and sustainable than they found it.

Of course, all of these practices are undermined by the effects of industry, such as the introduction of synthetic organisms or the patenting of traditional seeds by corporations so they can't be used by farmers in favor of genetically modified alternatives. Big Agra has even taken control of the criteria for organic certification in America, cynically prohibiting those who use the most regenerative farming practices from labeling their products organic. If the landscape is going to be defined by the law of "might makes right," then a patient, permaculture approach is going to be tricky to pull off.

And that war, in turn, entrenches food activists in an anti-technology corner, making them reluctant to incorporate the best of modern science into their operations. In many circumstances, sustainable practices, biodiversity, and livestock rotation can be augmented by synthetic nitrogen, solar power, or computerized irrigation. Progress isn't the enemy, so long as it's being used to embrace and support complexity, rather than attempting to eliminate it.

76.

The Japanese built a nuclear power plant right down the hill from the stone tablets that their ancestors put in the ground warning, "Don't build anything below here." The markers, called tsunami stones, were placed centuries ago by villagers who had experienced the region's devastating earthquakes and floods. Moderns ignored the advice, believing that their building techniques far surpassed anything their ancestors could have imagined.

The villagers had recognized the pattern of natural disasters, as well as the fact that the cycle repeated too infrequently for every generation to witness it. But their efforts to communicate their wisdom failed to impress a civilization without patience for pattern recognition or a sense of connection to the cyclical nature of our world.

Weather, ecology, markets, or karma: what goes around comes around. What the ancients understood experientially, we can today prove scientifically, with data and charts on everything from climate change to income disparity. But these facts seem not to matter to us unless they're connected to our moment-to-moment experience. Cold, abstract numbers carry the whiff of corrupt bureaucracy. With politicians actively undermining the importance of factual reality, NASA's climate data may as well be tsunami stones.

Phenomena such as climate change occur on time scales too large for most people to understand, whether they're being warned by scientists or their great-grandparents. Besides, the future is a distancing concept—someone else's problem. Brain studies reveal that we relate to our future self the way we relate to a completely different person. We don't identify that person as *us*. Perhaps this is a coping mechanism. If we truly ponder the horrific possibilities, everything becomes hyperbolic. We find it easier to imagine survival tactics for the zombie apocalypse than ideas to make the next ten years a bit better.

The endless succession of inspirational talks by well-meaning techno-solutionists with patented, world-saving ideas doesn't make the future feel any more real, either. Let's shoot reflective particles into the atmosphere, pour iron filings into the oceans, and dig pod tunnels for California commuters! These environmental moonshots put off sustainable abundance, as if it's something we can achieve only with yet more profitable high-tech innovation. The narratives all depend on linear, forward-moving, growth-based progress rather than the recognition of cycles or the retrieval of wisdom.

Such utopian projects make heroes out of the billionaires who envision them, while helping us justify our refusal to make any substantive changes to the way we live right now. If we bought fewer smartphones, we could enslave fewer children to produce them. If we ate less red meat, we would reduce our carbon footprint more than if we gave up our automobiles. Right now, today, we can do those things.

We don't need to build a network of solar-powered adobe homes on Mars. The future is not a discontinuity or some scenario we plan for so much as the reality we are creating through our choices right now. We just need to observe the flows, recognize the patterns, and apply them everywhere we can.

We can apply the regenerative principles of organic soil management to making the economy more circular. Just as we can derive an entire ethical framework from the single practice of veganism, we can apply the insights of permaculture practitioners to education, social justice, and government: look for larger patterns, learn from elders, understand and leverage natural cycles.

To begin, though, we have to touch ground. We humans need to become flow observers and pattern recognizers in our local realities and communities. Only then can we begin to have any sense of what is happening beyond our immediate experience, and forge solidarity with others.

77.

Science is not a cold abstraction, but a product of directly felt human experience.

If we think of science as the knowledge of nature, then it makes sense that its discoveries often come from those who are most intimately dependent on its processes: sailors, hunters, miners, healers, and others whose livelihoods involve direct encounters with nature's ways. Nearly every plant and animal species we eat is the result of selective breeding—a gentle form of genetic engineering, really—by working farmers long before Mendel founded the discipline of genetics. Our knowledge of the oceans and tides came from people whom Benjamin Franklin described as "simple whalers." Contemporary medicine still retrieves occasional insights from nontraditional sources.

Modern scientists are cautious not to romanticize the practices of healers and shamans, for along with any real science they may have unearthed came distinctly unscientific practices, from astrology to voodoo. By refusing to separate themselves from nature, the rationalists argue, these amateur practitioners were unable to achieve objectivity.

This view was best expressed in the 1600s by King James's

famed advisor, Francis Bacon. He believed that nature held secrets in her womb, and needed to be forcibly penetrated to make her give them up: "Nature must be taken by the forelock . . . lay hold of her and capture her . . . conquer and subdue her . . ." In the unrepentant language of a rape fantasy, he argued that we must see nature as a feminized object, rather than as a larger system of which we ourselves are part.

Science's great innovation—and limitation—was to break things down into their component parts. Science, from the root *sci-*, meaning "to split or cleave," dissects things in order to understand them. This makes sense: isolate a particular process, make a hypothesis about it, formulate an experiment, see if it produces repeatable results, and then share the knowledge with others. This is how we found out that objects have inertia, that sound has a speed, and that plants absorb CO_2.

These isolated, repeatable discoveries, in turn, make very particular things possible: for example, antibiotics—first used by ancient Egyptians but later refined in the lab—which turned lethal infections into minor annoyances. But to doctors armed with antibiotics, every problem started looking like a microbe. While antibiotics effectively killed their targets, they also killed helpful bacteria, compromised a patient's immune response, and encouraged the development of resistant strains. Worse, medical professionals were incentivized to develop ever more tightly focused medicines and treatments that could be monopolized for profit. The discovery that a common but unpatentable substance such as olive leaf extract has antiviral properties doesn't benefit the pharmaceutical industry any more than soil-enriching crop rotation benefits chemical companies.

Science must again become a holistic, human pursuit.

78.

Our common sense and felt experience contradict too much of what we're being told by scientific authorities. That's a problem. Research scientists' willingness to play along with industry and accept grants to prove the benefits of tobacco or corn syrup doesn't encourage us to place more trust in them either. If those arguing in favor of vaccination enjoyed more public credibility, for example, more people would see the logic and ethics of taking a minute risk in order to benefit our collective immunity.

Instead, we get a population increasingly distrustful of scientific evidence, whether it's about the low correlation between vaccines and autism or the high one between human activity and climate change. People fear that science misses the big picture, misrepresents reality to benefit its funders, or demands we follow counterintuitive and disempowering instructions.

The unemployed coal worker doesn't want to be retrained to build solar panels for a company thousands of miles away, owned by venture capitalists aligned with progressives screaming about climate change. He wants to create value the way his parents and grandparents did, by digging up the local resource right under his feet. Environmentalism feels like a cruel trick or an international conspiracy, and the patronizing tone of those who "know better" doesn't convince him otherwise.

By disconnecting science from the broader, systemwide realities of nature, human experience, and emotion, we rob it of its moral power. The problem is not that we aren't investing enough in scientific research or technological answers to our problems, but that we're looking to science for answers that ultimately require human moral intervention.

When science is used as a defense against nature rather than as a way of engaging more harmoniously with it, we disconnect ourselves from our moral core. We lose connection to the flows of vitality that animate the whole of life. This degrades the social, emotional, and ethical fabric from which we draw our strength, establish our priorities, and initiate positive change.

Just like corporatism, religion, and nationalism, science fell victim to a highly linear conception of the world. Everything is cause and effect, before and after, subject and object. This worked well for Newton and other observers of mechanical phenomena. They understood everything as having a beginning and an end, and the universe itself as a piece of graph paper extended out infinitely in all directions a background with absolute measure, against which all astronomical and earthly events take place.

Everything in material reality can be isolated and measured against these invented backgrounds, but the backgrounds don't actually exist. They're a convenient way for applied scientists to treat different parts and processes of the world as separate and independent. But they're not. There is no backdrop against which reality happens. An object doesn't sit anywhere absolute in space; its position is entirely a matter of its relation to every other object out there.

Like a dance where the only space that exists is defined by and between the dancers themselves, everything is happening in relationship to everything else. It's never over, it's never irrelevant, it's never somewhere else.

That's what forces science into the realm of morality, karma, circularity, and timelessness that prescientific people experienced. There's ultimately no ground on which a figure exists. It's all just ground, or all just figure. And humans are an inseparable part.

RENAISSANCE
NOW

79.

Built to enhance our essential interrelatedness, our digital networks could have changed everything. And the internet fostered a revolution, indeed. But it wasn't a renaissance.

Revolutionaries act as if they are destroying the old and starting something new. More often than not, however, these revolutions look more like Ferris wheels: the only thing that's truly revolving is the cast of characters at the top. The structure remains the same. So the digital revolution—however purely conceived—ultimately brought us a new crew of mostly male, white, libertarian technologists, who believed they were uniquely suited to create a set of universal rules for humans. But those rules—the rules of internet startups and venture capitalism—were really just the same old rules as before. And they supported the same sorts of inequalities, institutions, and cultural values.

A renaissance, on the other hand, is a retrieval of the old. Unlike a revolution, it makes no claim on the new. A renaissance is, as the word suggests, a rebirth of old ideas in a new context. That may sound less radical than revolutionary upheaval, but it offers a better way to advance our deepest human values.

The revolutionary fervor with which the digital era was promoted has finally begun to die down, and people are becoming aware of the ways in which these networks and the companies behind them have compromised our relationships, our values, and our thinking. This is opening us to the possibility that something much bigger is going on.

80.

Are we in the midst of a renaissance? Might the apparent calamity and dismay around us be less the symptoms of a society on the verge of collapse than those of one about to give birth? After all, childbirth is traumatic. Might we be misinterpreting a natural process as something lethal?

One way to evaluate the possibility would be to compare the leaps in art, science, and technology that occurred during the original Renaissance with those we're witnessing today. Do they have the same magnitude?

Perhaps the most dramatic artistic technique developed during the Renaissance was perspective painting. Artists learned how to render a three-dimensional image on a flat, two-dimensional canvas. What's our equivalent? Maybe the hologram, which lets us represent a fourth dimension of time on a flat plane. Or virtual reality, which lets the viewer experience a picture as an immersive environment.

During the Renaissance, European sailors learned to circumnavigate the globe, dispelling the conception of a flat earth and launching an era of territorial conquest. In the twentieth century we orbited and photographed our planet from space, launching a mindset of ecology and finite resources. The Renaissance saw the invention of the sonnet, a form of poetry that allowed for the first extended metaphors. We got hypertext, which allows anything to become a metaphor for anything else. The Renaissance got its own new media, too: the printing press, which distributed the written word to everyone. We got the computer and the internet, which distribute the power of publishing to everyone.

Most significantly, a renaissance asks us to take a dimensional leap: from flat to round, 2D to 3D, things to metaphors,

metaphors to hyperlinks, or top-down to peer-to-peer. The original Renaissance brought us from a flat world to one with perspective and depth. Our renaissance potentially brings us from a world of objects to one of connections and patterns. The world can be understood as a fractal, where each piece reflects the whole. Nothing can be isolated or externalized since it's always part of the larger system.

The parallels are abundant. This is our opportunity for renaissance.

01.

We might say that reason is to Reason as revolution is to renaissance. A renaissance without the retrieval of lost, essential values is just another revolution.

The first individuals and organizations to capitalize on the digital era ignored the underlying values that their innovations could have retrieved. They childishly assumed they were doing something absolutely new: disrupting existing hierarchies and replacing them with something or someone better—usually themselves. The early founders merely changed the ticker symbols on Wall Street from old tech companies to new tech companies, and the medium used to display them from paper tape to LEDs.

The digital revolution was no more than a superficial changing of the guard. Yet if we dispense with the need to believe our innovations are wholly original, we free ourselves to recognize the tidal patterns of which they are a part.

The original Renaissance, for instance, retrieved the values of ancient Greece and Rome. This was reflected not just in the philosophy, aesthetics, and architecture of the period, but in the social agenda. Central currency favored central authorities, nation-states, and colonialism. These values had been lost since the fall of the Roman Empire. The Renaissance retrieved those ideals through its monarchies, economics, colonialism, and applied science.

So what values can be retrieved by our renaissance? The values that were lost or repressed during the last one: environmentalism, women's rights, peer-to-peer economics, and localism. The over-rationalized, alienating approach to science is now joined by the newly retrieved approaches of holism and connectedness. We see peer-to-peer networks and crowdfunding replacing the top-down patronage of the Renaissance, retrieving a spirit of mutual aid and community. Even the styles and culture around this activity, from Burning Man and craft beer to piercing and herbal potions, retrieve the human-scaled, medieval sensibilities repressed by the Renaissance.

A renaissance does not mean a return to the past. We don't go back to the Middle Ages, bloodletting, feudalism, or sword fights in the street. Rather, we bring forward themes and values of previous ages and reinvent them in new forms. Retrieval makes progress less purely linear—not so much a ladder as a spiral staircase, continually repeating the same pattern, but ascending all the way. Retrieval helps us experience the insight of premodern cultures that nothing is absolutely new; everything is renewal.

Our general lack of awareness about the values being retrieved by digital technology made it easy for status quo powers to co-opt our renaissance and reduce it to just another revolution. So, like a counterculture packaged and resold to teens at the mall, new, potentially transformative threads are exploited by those looking to win the same old games. The 1960s' be-ins and free love communes were seized upon by lecherous men looking to leverage the openness of psychedelic culture for easy sex. The intellectual potential of the 1990s' internet was exploited by stock traders looking to sell new stories to investors. The unprecedented possibilities for connection offered by social media in the 2000s were surrendered to the more immediately profitable pursuits of surveillance, data mining, and user control. And the sharing economy of the 2010s was handily put down by venture capitalists, who used these same principles to establish incontestable and extractive platform monopolies.

Possibilities for renaissance are lost as our openness to fundamental change creates footholds for those who would exploit us. Innovations are instrumentalized in pursuit of short-term profit, and retrieved values are ignored or forcibly quashed.

Retrieval matters. Without retrieval, all our work and innovation is just research and development for the existing, repressive system. Tellingly, the commercial uses for a technology tend to emerge only *after* it has been around for a while. That's because they are not the reasons the technology was born in the first place.

Why is it so important to look back at what's being retrieved? Because retrieval is what connects us not just to the past but to core human motivations and values. This is, at heart, a positive, conservative impulse, because each time we bring forward a fundamental human value, we are ensuring that we bring ourselves—as humans—into the next environment.

82.

The most explicitly humanist value retrieved by the Renaissance, and the one we're most hampered by today, was the myth of the individual.

Leonardo da Vinci's iconic *Vitruvian Man*—the 1490 drawing of a man, perfectly proportioned within a circle and a square—presented the human form in the idealized geometric terms of the ancient Roman architect Vitruvius. The individual human body was celebrated as an analogy for the perfect workings of the universe.

Almost all of the period's innovations retrieved some aspect of individuality. The printing press—much to the frustration of the priests—gave everyone Bibles and the opportunity to interpret the gospel for themselves. This led to Protestantism, personal salvation, and a more individual relationship to God. Reading a book was a personal activity. The gentleman reading in his study retrieved the image of the Greek citizen: a slave-owning white male who thinks he lives by the principles of democracy.

Perspective painting similarly emphasized the point of view of an individual observer, as well as a single best position from which to view a work. Likewise, the plays and epic poems of the Renaissance retrieved the tragic hero of ancient Greek drama and poetry. The character of Dr. Faustus in Marlowe's play is often cited as the first to embody the Renaissance ideal of the individual. He is a self-made man, who has learned about the world through personal experimentation. But he also represents what happens when this commitment to selfhood goes too far: he makes a deal with the devil for total knowledge, and pursues his atomized self-interests over anything resembling social unity.

Marlowe was confronting something new to his society. It wasn't until the Renaissance that people started to think of themselves as having personal lives, struggles, and destinies—and that these individual interests had to be weighed against the public good. This gave rise to the Enlightenment, human rights, democracy—all almost unequivocally positive developments.

When the individual's needs are balanced with those of the

collective, things stay in balance. But when people are understood as self-interested individuals in competition against one another for survival and dominance, we get into trouble. Yet that's precisely what the economic reformations of the same Renaissance demanded. Central currency turned simple transactional tokens—a utility for exchange—into a scarce commodity. The Renaissance's chartered monopolies transformed craftspeople and cooperative commons users into expendable employees, competing for a day's work.

The myth of individuality made capitalism possible and has sustained it to this day. Economists modeled the market on the false premise that human beings are entirely rational individuals acting in their own self-interest. And corporations learned to stoke consumption by reinforcing our identities as individual consumers. Why sit on a streetcar with your friends when you can be driving your car, all alone? Why borrow your neighbor's lawnmower when you can have your own? What's a house in the suburbs without a fence defining your property?

Capitalism's vision of the individual as a completely self-interested being, whose survival was a Darwinian battle royale, is at odds with our social evolution and our neurobiology. But unless we consciously retrieve the power inherent in our collective nature, we will remain unable to defend ourselves against those who continue to use our misguided quest for individuality against us.

83.

The beauty of living in a renaissance moment is that we can retrieve what we lost the last time around. Just as medieval Europeans retrieved the ancient Greek conception of the individual, we can retrieve the medieval and ancient understandings of the collective. We can retrieve the approaches, behaviors, and institutions that promote our social coherence.

Revolution alone won't work. Neither will the blanket rejection of the values of the last renaissance, such as science, order, control, centrality, or even individuality. Instead, we should accept them as the context in which to bring forth their counterparts or, rather, complements. The ideals of the prior renaissance are the ground in which these lost ideals are retrieved.

We're moving from one understanding of our place in things to another. The Renaissance may have brought us from the tribal to the individual, but our current renaissance is bringing us from individualism to something else. We're discovering a collective sensibility that's more dimensional and participatory than the unconsciously formed communities of the past. We had to pass through this stage of individualism in order to get there.

Children must learn to separate from their parents and experience themselves as ego-driven individuals before they connect to others and establish meaningful relationships or intimacy. Likewise, human beings first needed to emerge from the ground and become figures: the subjects of our own stories. This was a dimensional leap—from one big blob to discrete individuals, each with our own intersectional identities. And now it's time for yet another leap.

84.

We don't yet have great ways for talking about this new spirit of collectivism. The relationship between individuals and society has always been framed as a necessary compromise: we are told we must sacrifice our personal goals for the sake of the many. But what if it's not a zero-sum, either/or? Humans, at our best, are capable of embracing seeming paradox. We push through the contradiction and find a dynamic sensibility on the other side.

We can think of our newly retrieved collectivism as a way of being both figure and ground at the same time. This is the idealized artistic community envisioned by Burning Man; it's the politics of consensus that informed Occupy; and it's the distributed economy aspired to by the open source and blockchain movements—to name just a few.

Each of these movements depends on our comfort with what we could call a fractal sensibility, the notion that each tiny part of a system echoes the shape and structure of the whole. Just as the veins within the leaf of a single fern reflect the branches, trees, and structure of an entire forest, the thoughts and intentions of a single individual reflect the consciousness of the whole human organism. The key to experiencing one's individuality is to perceive the way it is reflected in the whole and, in turn, resonate with something greater than oneself.

ORGANIZE

85.

Those of us seeking to retrieve some community and connection today do it with greater awareness of the alternatives. We don't retrieve collectivism by happenstance, but by choice. This enables us to consciously leverage the power of grassroots connections, bottom-up politics, and cooperative businesses—and build a society that is intentionally resilient and resistant to the forces that would conquer us.

Early Internet enthusiasts had little understanding of how the network's anticommercial safeguards protected the humanistic values of its fledgling culture. Rave kids didn't understand the real power of their ecstatic rituals, which was to reclaim the public spaces in which to hold them. Many members of Occupy were too dismayed by losing Zuccotti Park to realize that the bigger victory was to develop a new normative behavior for activists and a consensus-driven approach to administrating the demos.

Likewise, the members of early communal groups were generally unaware of the real power of solidarity. They weren't collectivist by design so much as by nature. Members of a group just tended to believe the same things. Their labor wasn't particularly specialized, either. They were necessarily bound by their common location, needs, and worldviews.

Only after humans emerged as individuals, with differentiated perspectives, conflicting beliefs, specialized skills, and competing needs could we possibly comprehend collectivism as an active choice. It is in that positive determination to be members of Team Human that we derive the power and facility to take a deliberate stand on our own behalf.

86.

Solidarity begins with place.

While it's tempting to rally around the emotionally charged issues of the mainstream media, they tend to be abstract, polarizing, and unconnected to people's lived experience. Whether or not an enemy politician is indicted for deleting the wrong email has a whole lot less impact on real people than whether there are chemicals leaching into the water supply, enough funds for the schools, or places for the elderly to go during the day.

When politics are truly local, issues can't be so easily distorted by the partisan framing of cable television. One remote county's redistricting of farmland or revised policy on recycling doesn't interest national news conglomerates looking for ratings, anyway. And on the rare occasions when a local issue does show up on the national news, those of us with direct knowledge of the story can only marvel at how our reality differs from what's being depicted on the screen. All national news is distorted—we just have no way of evaluating it.

Team Human participates in national and global politics from the bottom up, the small to the large, and the local to the national and beyond. We can still be guided by bigger principles, but those principles are informed by living in a community, not by listening to talk radio.

It's not easy. Local debate on almost any issue ends up being more challenging than most of us expect, but even the most contentious town hall conflicts are tempered by the knowledge that we have to live together, in peace, after the fight is over. When you live somewhere, you can't just tune to a different channel and get different neighbors. Everything comes back around.

DOUGLAS RUSHKOFF

87.

Global relations are forged locally, as well. The most important part of any diplomatic journey is the last foot—the one that brings the two parties face to face. For it is there in the live encounter that potential adversaries are forced to recognize each other's humanity.

This is the theory of diplomacy that led to the famous Oslo Accords between the warring parties of the Middle East—an agreement that failed only because one of the signatories was assassinated by an extremist among his own people. Such sociopathic behavior is not unusual for religious fanatics. Anyone who has become so distanced from other people that they see humans as less important than their ideology will act in anti-human ways.

Citizen diplomacy—just tourism, really, where a nation's people represent its values abroad—has long been recognized as the most productive tool for improving international relations. Propaganda is manipulative. It begets competition between those who seek to dominate public opinion. Citizen diplomacy, on the other hand, is behavioral: showing by example, live and in person. Instead of leading to confrontation, it engenders interdependence.

Whether between allies or adversaries, voters or towns-people, face-to-face engagement leverages our evolved capacity to establish rapport. While our minds may be determined to win our agenda, our hearts just want to win over the other person. If we're capable of engaging in a genuine conversation, our common agenda as humans far outweighs the political platforms we've signed onto. This is not weakness, but strength.

88.

Each of us can't do everything. Representative democracy gives us the chance to choose other people to speak on our behalf—ideally, in face-to-face interactions with representatives of other stakeholders.

By relegating the democratic process to behemoth media and internet companies, we dispense with both the power of rapport and the connection to place. This makes us more likely to see one another as less than human, and act or vote inhumanely ourselves.

In repeated experiments, social media platforms have demonstrated their ability to induce people to vote or not vote by placing particular messages in their news feeds. Combine this with the same companies' ability to predict how someone is likely to vote and we get a powerful tool for undermining democracy by manipulating voters' behaviors. No fake news required.

We have no way of knowing when this is being done to us, or what other techniques are being used. Social media platforms have no facial expressions through which we can detect their duplicity. They see us, but we can't see them. In these virtual spaces, we can't even truly engage with the faces of other people. Our impulse for mutual aid and human connection remains dormant.

Denied this contact, we start to feel alone and angry, making us easy targets for any demagogue who wants to stoke our rage and trigger our most sociopathic tendencies. The other side becomes an inferior species. Losers.

There's nothing wrong with opposing someone. But our encounters with our adversaries must be grounded in the greater context of our shared humanity. This means that in every encounter, the human-to-human, I-and-you engagement itself becomes the main event.

The other person's position—even a heinous one—still derives from some human sensibility, however distorted by time, greed, war, or oppression. To find that core humanity, resonate with it, and retrieve its essential truth, we have to be willing to listen to our adversaries as if they were human.

They are human—at least for now.

89.

Does everyone get to be on Team Human? How do we practice social inclusion with those who don't want to be included?

If we call summon enough of our own humanity to really listen, we'll find that most of our counterparts are not truly violent or irredeemably mean-spirited. Understanding their fears and then working together toward our common goals is far more constructive than pretending that entire populations have no humanity at all. This means venturing as far back into their emotional logic as it takes to find something we can identify with—the valid feelings they hold, before they are transformed into their more destructive expressions.

The enemies of a tolerant, inclusive culture don't see their position as inhumane or racist. They see the history of the world as a competition for dominance, and their own race and civilization as the rightful, if unrecognized, winner. What some of us understand as the aggressive, colonial expansion of white European nations, they see as the growth and spread of civilization itself. Might makes right. So, in their view, those of us who are attempting to promote the cultures, values, and languages of defeated peoples are retrieving failed approaches, teaching them to our kids, and weakening our society. These cultures don't deserve to get taught, they feel, because they *lost*.

The Silicon Valley billionaire with an apocalypse bunker in New Zealand uses a similar logic to justify creating the very conditions that are leading to a world where such a plan B should be required. The smartest, wealthiest technologist gets to survive because he won. It's a hyperbolic, digitally amplified, zero-sum version of the same exclusion.

So how do we engage with such a perspective? First, by acknowledging the vulnerability for which it is attempting to compensate. Free-market fanatics, racist bullies, and techno-elites all see nature as a cruel competition. A jungle. For all of nature's cooperative networks, there are just as many species hunting and eating others. Letting down one's guard means being vulnerable to attack.

Once we really see where they are coming from, we can empathize with their fear, follow them into that dark scary place, and then find a better way out *with* them. We give our potential antagonists a solution other than violence or retreat: Yes, nature is a potentially cruel place, but it's not aware of that cruelty. The cheetah isn't intending to be cruel when it attacks the gazelle anymore than the tornado intends to be cruel to the town it destroys. Only humans are conscious enough to see nature's actions as injustice; we respond by assisting those who have been impacted and taking measures to protect them in the future. This is, quite possibly, our unique ability and purpose as a species. We humans are nature's conscience, and we do what is in our power to make nature more, well, humane.

The extent to which we can do this is a sign not of weakness, but of strength. The leader who must attack another people to demonstrate dominance is weaker than the one who is confident enough to let them live in peace. The government that must surveil and control its citizens is less secure in its authority than the one that sets them free. The technologists working on apocalypse bunkers are not as confident in their talents as the ones working to prevent whatever calamities they fear. The people who exclude others based on race or class are less secure in their own legitimacy and competence than those who welcome everyone.

Similarly, the person who can't see a human being behind the mask of intolerance is weaker than the one who can.

90.

The people with whom we disagree are not the real problem here. The greatest threats to Team Human are the beliefs, forces, and institutions that separate us from one another and the natural world of which we are a part. Our new renaissance must retrieve whatever helps us reconnect to people and places.

The largest organic association of people is the city. Organized around resources, the commons, and marketplaces, cities grow from the bottom up. As a natural amalgamation of humans, they take on the qualities of any collective organism, such as a coral reef or a termite mound.

During the Renaissance, the primacy of the city-state of people was surrendered to the politically determined nation-state, an invented concept. The transition from cities to nations transformed people from members of a community to citizens of a state. Localities were disempowered—along with their currencies and circular economies—as resources and attention were directed upward, first to monarchs and then to corporations.

Our local peer-to-peer interactions, solidarity, and collective concerns were replaced by a large-scale, abstracted democratic process that couldn't help but become more like the expression of brand affinities than of human needs. We became individuals voting our personal preferences in the seclusion of a booth, rather than groups expressing their solidarity through collaboration.

This same process continues to this day. What the British East India Company, European colonial empires, and modern transnational corporations did to cooperative peoples of the world, today's digital companies are doing to us: disconnecting us from the ground on which we stand, the communities where we live, and the people with whom we conspire. To conspire literally means to "breathe together"—something that any group of people meeting together in real space is already doing.

This is why we must reclaim terra firma, the city, and the physical communities where this can happen. Humans derive their power from place. We are the natives here, and we have the home field advantage.

91.

Maintaining home field advantage means staying in the real world. But sometimes it's hard to know what's really real.

We must learn to distinguish between the natural world and the many constructions we now mistake for preexisting conditions of the universe. Money, debt, jobs, slavery, countries, race, corporatism, stock markets, brands, religions, government, and taxes are all human inventions. We made them up, but we now act as if they're unchangeable laws. Playing for Team Human means being capable of distinguishing between what we can't change and what we can.

It's akin to the stages that video game players pass through on their way to mastery. Kids begin as pure players, tacitly accepting the rules of the game. They may begin the game without even reading the rules, and then proceed as far as they can on their own.

What do they do when they eventually get stuck? Some simply give up, or become the spectators of professional gamer shows. But those who want to proceed to the next level of mastery go online and find strategies, play-throughs, and "cheat codes." These are tricks that give the player infinite ammunition or the extra shield strength they need to get more easily to the next levels. So the player has progressed from being a mere player to a cheater—someone playing from outside the original confines of the game. It's not until this stage that the player even has the option of moving beyond the original rules.

For players who get to the end and still want more, many games offer the ability to "mod" their own versions. The players can build new levels with new obstacles, or even remake the dungeons of a castle into the corridors of a high school. The players moves from being a cheater to an author. But they must still accept the basic premise of the game and its operating system.

They upload their modified levels to game sites, where they are downloaded by other enthusiasts. Those who have designed the most popular levels may even be asked to work at game companies, designing entirely new games. These kids have gotten beyond the author stage to that of programmer. They're capable of writing new laws.

These stages of video game play—from player to cheater to author to programmer—are analogous to the stages we passed through as a civilization. Before the written word, we simply received law. After we could read, we had the power of interpretation. With printing came the ability to share our opinions. With programming comes the opportunity to write new rules.

If nothing else, living in a digital media environment should help us become more aware of the programming all around us. We should be able to recognize the suburbs as an experiment in social control through isolation, and corporatism as an effort to stunt bottom-up prosperity. Simply remembering that corporations were invented should alone empower us to reinvent them to our liking.

Team Human can organize, take to the streets, participate in electoral politics, develop new platforms for discussion, engage more purposefully with the natural world, and work to reform corrupt institutions or to build better ones.

This, not new software, is the underlying empowerment to be seized in a digital age. The original digits are the fingers on our own human hands. By retrieving the digital, we take a hands-on approach to remaking the world in our own and everyone's best interests.

We participate when we can, and change the rules when we can't.

YOU ARE
NOT ALONE

92.

Human beings can intervene in the machine. That's not a refusal to accept progress. It's simply a refusal to accept any particular outcome as inevitable.

Team Human doesn't reject technology. Artificial intelligence, cloning, genetic engineering, virtual reality, robots, nanotechnology, bio-hacking, space colonization, and autonomous machines are all likely coming, one way or another. But we must take a stand and insist that human values are folded into the development of each and every one of them.

Those of us who remember what life was like before these inventions remade the landscape have a duty to recall and proclaim the values that are being left behind.

93.

Our values—ideals such as love, connection, justice, distributed prosperity—are not abstract. We have simply lost our capacity to relate to them.

Values once gave human society meaning and direction. Now this function is fulfilled by data, and our great ideals are reduced to memes. Instead of consciously reenacting the essential, mythical truths of our existence, we track aggregate opinion and stoke whatever appetites guarantee the greatest profit.

A consumer democracy cannot express our higher values as humans; it amounts to little more than metrics. Neither the sales of the latest smartphone nor the votes for the latest autocrat prove the real merits of either.

We need another way to express and execute the human agenda.

94.

The programmer necessarily sees the world as suffering from bad software: better coding makes for happier people. But such tactical approaches address only granular problems. Instrumentalism solves a problem without the benefit of the greater human context. It maximizes efficiency in the short run, while externalizing its costs to someone or someplace else.

Achieving a higher human value such as universal justice is not a question of engineering. Blockchains and robots don't address the fundamental problem of humanity's widespread refusal to value one another and the world we share.

Rather than trying to locate our values in better computer code, we should be turning to the parts of ourselves that can't be understood in those terms: our abilities as humans to engage with ambiguity, to retrieve the essential, and to play well with others.

95.

A value-driven society is like a self-governed or "holocratic" organization. Without a leader to follow or lines of code to direct us, we must instead be driven by a set of common ideals.

These sorts of ideals can't be justified with numbers or instrumentalized with technology. They emerge from a sense of shared ownership and responsibility for our collective interest.

Co-ops help engender this sensibility, where all the workers are also stakeholders in the enterprise. They are a group of autonomous individuals working toward shared goals and holding complementary worldviews. The common values are what establish the direction of the group's energy, and serve as the moral compass for collective decision-making.

The co-op is also a model for a society of autonomous, collaborating grownups. Communities composed of genuinely interdependent individuals, in which each of us is both needed and accountable, offer the highest quality of life and greatest level of happiness. Our personal contributions have greater effect when they are amplified by a network of peers working in solidarity. The individual is actualized through the community.

Only service to others gives us the opportunity to experience autonomy and belonging at the same time.

96.

Humans still like to compete. We don't have to be absolutely generous all the time. There's a place for aggression and entrepreneurialism, winners and losers. It just has to happen, like sports, with rules and transparency.

A humane civilization learns to conduct its competitive activities within the greater context of the commons. Our courts, democracy, markets, and science are all characterized by competition, but this competition takes place on highly regulated playing fields. The free market is not a free-for-all, at all, but a managed game with rules, banks, tokens, patents, and stock shares.

These competitive spaces only work in everyone's long-term interest when their operations are radically transparent. Courts can't provide justice if we don't know why certain people are jailed and others are not. Markets don't function when certain players get information that others don't. Science can't advance if results are not shared and verifiable.

Transparency is really the only choice, anyway. We can't hide from and lie to one another any longer. There's just no point to it. If a stage magician can read our faces and detect our false statements, then we all must know on some level when we're being deceived by another person. When our media and machines are both opaque and untrustworthy, we must learn to depend on one another for the truth of what is going on here.

If we want to steer our society back toward reality, we have to stop making stuff up.

97.

Things may look bleak, but the future is open and up for invention.

We mistakenly treat the future as something to prepare for. Companies and governments hire scenario planners to lay out the future landscape as if it were a static phenomenon. The best they can hope for is to be ready for what is going to happen.

But the future is not something we arrive at so much as something we create through our actions in the present. Even the weather, at this point, is subject to the choices we make today about energy, consumption, and waste.

The future is less a noun than a verb, a thing we do. We can futurize manipulatively, keeping people distracted from their power in the present and their connection to the past. This alienates people from both their history and their core values.

Or we can use the idea of the future more constructively, as an exercise in the creation and transmission of values over time. This is the role of storytelling, aesthetics, song, and poetry.

Art and culture give us a way to retrieve our lost ideals, actively connect to others, travel in time, communicate beyond words, and practice the hard work of participatory reality creation.

98.

There's only one thing going on here.

As much as we think we're separate individuals, we're wired from birth and before to share, bond, learn from, and even heal one another. We humans are all part of the same collective nervous system. This is not a religious conviction but an increasingly accepted biological fact.

We can't go it alone, even if we wanted to. The only way to heal is by connecting to someone else.

But it also means that when one of us is disturbed, confused, violent, or oppressed, the rest of us are, too. We can't leave anyone behind or none of us really makes it to wherever we think we're going. And we can't just stay confused and depressed ourselves without confusing and depressing everyone who is connected to us.

This is a team sport.

99.

You are not alone. None of us are.

The sooner we stop hiding in plain sight, the sooner we can avail ourselves of one another. But we have to stand up and be seen. However imperfect and quirky and incomplete we may feel, it's time we declare ourselves members of Team Human.

On being called by God, the biblical prophets would respond "*Hineni*," meaning, "I am here." Scholars have long debated why

a person should have to tell God they're present. Surely they know God sees them.

Of course, the real purpose of shouting "*Hineni*" is to declare one's readiness: the willingness to step up and be a part of the great project. To call out into the darkness for others to find us: *Here I am.*

It's time for us to rise to the occasion of our own humanity. We are not perfect, by any means. But we are not alone. We are Team Human.

100.

Find the others.

NOTES

Notes are organized according to numbered sections of the text.

3.

People who are disconnected from the organizations or communities they serve often wither without them
    Jeffrey L. Metzner and Jamie Fellner, "Solitary Confinement and Mental Illness in U.S. Prisons: A Challenge for Medical Ethics," *Journal of the American Academy of Psychiatry and the Law* 30, no. 1 (March 2010).

8.

an invisible landscape of mushrooms and other fungi connecting the root systems of trees in a healthy forest
    Suzanne Simard, "How Trees Talk to Each Other," TED talk, June, 2016.

When the leaves of acacia trees come in contact with the saliva of a giraffe, they release a warning chemical
    Peter Wohlleben, *The Hidden Life of Trees: What They Feel, How They Communicate* (Vancouver: Greystone, 2016).

9.

"Individualists" who challenged the leader's authority or wandered away
    Merlin Donald, *Origins of the Modern Mind: Three Stages in the Evolution of Culture and Cognition* (Cambridge, MA: Harvard University Press, 1991).

The virtual combat benefits not just the one who would be killed
    Laszlo Mero, *Moral Calculations: Game Theory, Logic, and Human Frailty* (New York: Springer Science + Business, 1998).

John Marzluff and Russel P. Balda, *The Pinyon Jay: Behavioral Ecology of a Colonial and Cooperative Corvid* (Cambridge, UK: Academic Press, 1992).

10.

The most direct benefit of more neurons and connections in our brains is an increase in the size of the social networks we can form

Norman Doidge, *The Brain That Changes Itself* (New York: Penguin, 2007).

Developing bigger brains allowed human beings to maintain a whopping 150 stable relationships at a time

Robin Dunbar, *Human Evolution: Our Brains and Behavior* (New York: Oxford University Press, 2016).

11.

Social losses such as the death of a loved one, divorce, or expulsion from a social group, are experienced as acutely as a broken leg

Matthew D. Lieberman, *Social: Why Our Brains Are Wired to Connect* (New York: Crown, 2013).

Managing social relationships also required humans to develop what anthropologists call a "theory of mind"

Leslie C. Aiello and R. I. M. Dunbar, "Neocortex Size, Group Size, and the Evolution of Language," *Current Anthropology* 34, no. 2 (April 1993).

Prosocial behaviors such as simple imitation—what's known as mimesis—make people feel more accepted and included

Robert M. Seyfarth and Dorothy L. Cheney, "Affiliation, empathy, and the origins of theory of mind," *Proceedings of the National Academy of Sciences of the United States of America* 110 (Supplement 2) (June 18, 2013).

In one experiment, people who were subtly imitated by a group produced less stress hormone

Marina Kouzakova et al., "Lack of behavioral imitation in human interactions enhances salivary cortisol levels," *Hormones and Behavior* 57, no. 4–5 (April 2010).

Our bodies are adapted to seek and enjoy being mimicked

S. Kuhn et al., "Why do I like you when you behave like me? Neural mechanisms mediating positive consequences of observing someone being imitated," *Social Neuroscience* 5, no. 4 (2010).

We naturally try to resonate with the brain state of the crowd

Thomas Lewis, Fari Amini, and Richard Lannon, *A General Theory of Love* (New York: Knopf, 2001).

12.

Humans are defined not by our superior hunting ability so much as by our capacity to communicate, trust, and share

Glynn Isaac, "The Food-Sharing Behavior of Protohuman Hominids," *Scientific American*, April 1978.

But contemporary research strongly supports more generous motives in altruism, which have nothing to do with self-interest

The Evolution Institute, https://evolution-institute.org.

It was a dangerous adaptation that involved crossing the airway with the foodway

Merlin Donald, *Origins of the Modern Mind: Three Stages in the Evolution of Culture and Cognition* (Cambridge, MA: Harvard University Press, 1991).

13.

The difference between plants, animals, and humans comes down to what each life form can store, leverage or—as this concept has been named—"bind"
Alfred Korzybski, *Science and Sanity: An Introduction to Non-Aristotelian Systems and General Semantics* (New York: Institute of General Semantics, 1994).

14.

Happiness is not a function of one's individual experience or choice, but a property of groups of people
Nicholas A. Christakis and James H. Fowler, "Dynamic spread of happiness in a large social network: Longitudinal analysis over 20 years in the Framingham Heart Study," *British Medical Journal* (December 4, 2008).

Young men with few social acquaintances develop high adrenaline levels
Sarah Knox, Töres Theorell, J. C. Svensson, and D. Waller, "The relation of social support and working environment to medical variables associated with elevated blood pressure in young males: A structural model," *Social Science and Medicine* 21, no. 5 (1985).

Lonely students have low levels of immune cells
J. K. Kiecolt-Glaser et al., "Marital quality, marital disruption, and immune function" *Psychosomatic Medicine* 49, no. 1 (January 1987).

15.

Making the independent choice to trust other people, or even to engage in self-sacrifice, allows people to feel that they are connected to a bigger project
Chris Hedges, "Diseases of Despair," *TruthDig*, September 3, 2017.

We engage in futile gestures of permanence, from acquiring wealth to controlling other people
Ernest Becker, *The Denial of Death* (New York: Free Press, 1977).

Mental health has been defined as "the capacity both for autonomous expansion and for homonomous integration with others"

Andras Angyal, *Neurosis and Treatment: A Holistic Theory* (Hoboken: John Wiley and Sons, 1965).

18.

Speech created a way of actively misrepresenting reality to others

Robert K. Logan, *The Extended Mind: The Emergence of Language, the Human Mind and Culture* (Toronto: University of Toronto Press, 2007).

When once writing appeared, it was accompanied by war and slavery

John Lanchester, "The Case Against Civilization," *New Yorker*, September 18, 2017.

For all the benefits of the written word, it is also responsible for replacing an embodied, experiential culture with an abstract, administrative one

Walter Ong, *Orality and Literacy* (London: Routledge, 1982).

Leonard Shlain, *The Alphabet Versus the Goddess* (London: Penguin, 1999).

marketing psychologists saw in [television] a way to mirror a consumer's mind

W. R. Simmons, "Strangers into Customers," marketing study prepared for National Broadcasting Co., New York, 1954.

Television told people they could choose their own identities

David Halberstam, *The Fifties* (New York: Ballantine, 1993).

The Century of the Self, film, directed by Adam Curtis (2005; United Kingdom: BBC Two, RDF Television).

Stuart Ewen, *All-Consuming Images* (New York: Basic Books, 1990).

Television was widely credited as the single biggest contributor to the desocialization of the American landscape

Robert D. Putnam, *Bowling Alone: The Collapse and Revival of American Community* (New York: Simon and Schuster, 2000).

19.

Gone was connectivity between people, replaced by "one-to-one marketing" relationships

Don Peppers and Martha Rogers, *The One to One Future* (New York: Currency, 1993).

We announced that the net was and would always be a "social medium"

Douglas Rushkoff, "The People's Net," *Yahoo Internet Life*, July 2001.

21.

The term "media virus"

Douglas Rushkoff, *Media Virus!* (New York: Ballantine, 1994).

23.

Memetics, the study of how memes spread and replicate, was first popularized by an evolutionary biologist in the 1970s

Richard Dawkins, *The Selfish Gene* (Oxford: Oxford University Press, 1976).

It's why a tame grasshopper

Brigid Hains, "Die, Selfish Gene, Die," *aeon*, December 13, 2013.

24.

they invest in propaganda from all sides of the political spectrum

Nancy Scola and Ashley Gold, "Facebook, Twitter: Russian Actors Sought to Undermine Trump After Election," *Politico*, October 31, 2017.

25.

The idea of figure and ground was first posited by a Danish psychologist in the early 1900s

Jörgen L. Pind, *Edgar Rubin and Psychology in Denmark* (Berlin: Springer, 2014).

26.

A good education was also a requirement for a functioning democracy

John Dewey, *Democracy and Education* (New York: Free Press, 1997).

The Soviets' launch of the Sputnik satellite in the 1960s led America to begin offering advanced math in high school

Alvin Powell, "How Sputnik Changed U.S. Education," *Harvard Gazette*, October 11, 2007.

27.

For these reasons, many of the most ambitious engineers, developers, and entrepreneurs end up dropping out of college altogether

Bill Gates, Steve Jobs, Mark Zuckerberg, Evan Williams, Travis Kalanick, Larry Ellison, Michael Dell, John Mackey, Jan Koum, to name only a few.

28.

Consider Thomas Jefferson's famous invention, the dumbwaiter

Silvio A. Bedini, *Thomas Jefferson: Statesman of Science* (Basingstoke: Palgrave–MacMillan, 1990).

Even today, Chinese laborers "finish" smartphones by wiping off any fingerprints

Victoria Turk, "China's Workers Need Help to Fight Factories' Toxic Practices," *New Scientist*, March 22, 2017.

29.

The Muzak in the supermarket is programmed to increase the rhythm at which we place things into our shopping carts

Douglas Rushkoff, *Coercion* (New York: Riverhead, 2000).

Our technologies change from being the tools humans use into the environments in which humans function

David M. Berry and Michael Dieter, *Postdigital Aesthetics: Art, Computation, and Design* (Basingstoke: Palgrave–MacMillan, 2000).

Think of the way video game graphics advanced

Heather Chaplin, *Smartbomb: The Quest for Art, Entertainment, and Big Bucks in the Video Game Revolution* (Chapel Hill: Algonquin, 2006).

30.

As the number of amphetamine prescriptions for young people continues to double

"Prescribed Stimulant Use for ADHD Continues to Rise Steadily," National Institute of Mental Health, press release, September 28, 2011, https://www.nih.gov/news-events/news-releases/prescribed-stimulant-use-adhd-continues-rise-steadily.

32.

The networks compressed distance

See the Nettime email list from the early 1990s, at https://www.nettime.org/archives.php, or Usenet groups such as alt.culture.

Even the corporate search and social platforms that later came to monopolize the net originally vowed never to allow advertising

Benoit Denizet-Lewis, "Why Are More American Teenagers Than Ever Suffering from Severe Anxiety?" *New York Times Magazine*, October 15, 2017.

33.

Persuasive technology, as it's now called, is a design philosophy taught and developed at some of America's leading universities
The current leader in the field is BJ Fogg's Captology Laboratory at Stanford University.

On the contrary, people change their attitudes to match their behaviors
"BDI Behaviour Change," Behavioural Dynamics Institute, 2014, https://www.youtube.com/watch?v=l3k_-k1Mb3c.

the sort of psychological manipulation exercised in prisons, casinos, and shopping malls
Douglas Rushkoff, *Coercion* (New York: Riverhead, 2000).

That's why academic studies of slot machine patterning became required reading
Natasha Dow Schull, *Addiction by Design: Machine Gambling in Las Vegas* (Princeton: Princeton University Press, 2014).
Nir Eyal, *Hooked: How to Build Habit-Forming Products* (New York: Portfolio, 2014).

the play is often surrendered to some pretty unplayful outcomes
Brian Burke, *Gamify: How Gamification Motivates People to Do Extraordinary Things* (Abingdon, UK: Routledge, 2014).
Kevin Werbach, *For the Win: How Game Thinking Can Revolutionize Your Business* (Philadelphia: Wharton Digital Press, 2012).
Jane McGonigal, *Reality Is Broken: Why Games Make Us Better and How They Can Change the World* (London: Penguin, 2011).

34.

We now know, beyond any doubt, that we are dumber when we are using smartphones and social media
Debra Kaufman, "Studies Show Smartphones, Social Media Cause Brain Drain," *etcentric*, October 10, 2017.

35.

human beings require input from organic, three-dimensional space
William Softky, "Sensory Metrics of Neuromechanical Trust,"
Journal of Neural Computation 29, no. 9 (September 2017).

We remember things better when we can relate them to their physical locations
Luke Dittrich, *Patient H.M.: A Story of Memory, Madness, and Family Secrets* (New York: Random House, 2017).

Our relationships become about metrics, judgments, and power
Benoit Denizet-Lewis, "Why Are More American Teenagers Than Ever Suffering from Severe Anxiety?" *New York Times Magazine*, October 11, 2017.

36.

Surprisingly, the inability to establish trust in digital environments doesn't deter us from using them
William Softky, "Sensory Metrics of Neuromechanical Trust,"
Journal of Neural Computation 29, no. 9 (September 2017).

This is how products you may have looked at on one website magically show up as advertisements on the next.
Do Not Track, documentary film, directed by Brett Gaylor (2015), available at https://donottrack-doc.com.

37.

We described ourselves as living in a "clockwork universe"
John of Sacrobosco, *De Sphaera Mundi (On the Sphere of the World)*, c. 1230, available at http://www.esotericarchives.com/solomon/sphere.htm.

Dennis Des Chene, *Spirits and Clocks: Machine and Organism in Descartes* (Ithaca, NY: Cornell University Press, 2000).

We sought to operate faster

George Lakoff, *Metaphors We Live By* (Chicago: University of Chicago Press, 1980).

Lewis Mumford, *Myth of the Machine* (Boston: Mariner, 1971).

Neil Postman, *Technopoly: The Surrender of Culture to Technology* (New York: Vintage, 1993).

Jean Baudrillard, *Simulacra and Simulation* (Ann Arbor: University of Michigan Press, 1994).

It's not just treating machines as living humans; it's treating humans as machines

John Seely Brown and Paul Duguid, *The Social Life of Information* (Cambridge, MA: Harvard Business Review Press, 2000).

38.

Study after study has shown that human beings cannot multitask

Clifford Nass, "Cognitive Control in Media Multitaskers," *Proceedings of the National Academy of Sciences* 106, no. 27 (September 15, 2009).

39.

The digital media environment expresses itself in the physical environment

Richard Maxwell and Toby Miller, *Greening the Media* (Oxford: Oxford University Press, 2012).

40.

National governments were declared extinct

John Perry Barlow, "Declaration of Independence of Cyberspace," *Wired*, February 8, 1996.

We are not advancing toward some new, totally inclusive global society, but retreating back to nativism

Marshall McLuhan, *Understanding Media* (Cambridge, MA: MIT Press, 1994).

Even 9/11 was a simultaneously experienced, global event
Jean-Marie Colombani, "Nous Sommes Tous Américains," *Le Monde*, September 12, 2001.

At the height of the television media era, an American president
Ronald Reagan, "Tear Down This Wall!" speech, June 12, 1987.

demand the construction of walls
Donald Trump, speech, Phoenix, August 31, 2016.

41.

In 1945, when Vannevar Bush imagined the "memex," on which computers were based
Vannevar Bush, "As We May Think," *The Atlantic*, July 1945.

Similar tensions are rising in India, Malaysia, and Sudan
Kevin Roose, "Forget Washington. Facebook's Problems Ahead Are Far More Disturbing," *Washington Post*, October 29, 2017.

42.

Highways divided neighborhoods, particularly when they reinforced racial and class divisions
Douglas Rushkoff, *Life, Inc.: How the World Became a Corporation and How to Take It Back* (New York: Random House, 2011).

New studies on the health effects of cellphones and WiFi
Dina Fine Maron, "Major Cell Phone Radiation Study Reignites Cancer Questions," *Scientific American*, May 27, 2016.
Jeneen Interlandi, "Does Cell Phone Use Cause Brain Cancer? What the New Study Means For You," *Consumer Reports*, May 27, 2016.

As if to celebrate their commitment to digital values, some school principals
Spike C. Cook, Jessica Johnson, and Theresa Stager, *Breaking Out of Isolation: Becoming a Connected School Leader* (Thousand Oaks, CA: Corwin, 2015).

Lisa Dabbs and Nicol R. Howard, *Standing in the Gap: Empowering New Teachers Through Connected Resources* (Thousand Oaks, CA: Corwin, 2015).

Ringo Starr, the Beatles drummer, famously lagged
Zumic Staff, "Great Drummers Break Down Ringo Starr's Style with the Rock and Roll Hall of Fame," *Zumic*, July 8, 2015.

so close to the "normal" beat of the song that it would be immediately corrected
Stephen Bartolomei, "Silencing Music, Quieting Dissent: How Digital Recording Technology Enforces Conformity Through Embedded Systems of Commodification," master's thesis, Queens College, City University of New York, 2016.

43.

transhumanists hope to ease our eventual, inevitable transition to life on a silicon chip
Julian Huxley, *New Bottles, New Wine* (New York: Harper Brothers, 1957).

The cycles of life are understood not as opportunities to learn or let go, but as inconveniences to ignore or overcome
Steven Salzberg, "Did a Biotech CEO Reverse Her Own Aging Process? Probably Not," *Forbes*, August 1, 2016.

44.

Or they sell the printers at a loss and then overcharge us for the ink cartridges
Chris Hoffman, "Why Is Printer Ink So Expensive?" *How-To Geek*, September 22, 2016.

45.

Technology is not driving itself
Kevin Kelly, *What Technology Wants* (London: Penguin, 2011).

They worked for themselves, fewer days per week, with greater profits, and in better health

Juliet B. Schor, *The Overworked American: The Unexpected Decline of Leisure* (New York: Basic Books, 1993).

They came up with two main innovations

Douglas Rushkoff, *Life, Inc.: How the World Became a Corporation and How to Take It Back* (New York: Random House, 2011).

46.

human beings now strive to brand themselves in the style of corporations

Taylor Holden, "Give Me Liberty or Give Me Corporate Personhood," *Harvard Law and Policy Review*, November 13, 2017.

the New York Stock Exchange was actually purchased by its derivatives exchange in 2013

Nina Mehta and Nandini Sukumar, "Intercontinental Exchange to Acquire NYSE for $8.2 Billion," *Bloomberg*, December 20, 2012.

47.

digital technology came to the rescue, providing virtual territory for capital's expansion

Joel Hyatt, Peter Leyden, and Peter Schwartz, *The Long Boom: A Vision for the Coming Age of Prosperity* (New York: Basic Books, 2000).

Kevin Kelly, *New Rules for a New Economy* (London: Penguin, 1999).

corporate returns on assets have been steadily declining for over seventy-five years

John Hagel et al., foreword, *The Shift Index 2013: The 2013 Shift Index Series* (New York: Deloitte, 2013).

48.

One or two superstars get all the plays, and everyone else sells almost nothing

M. J. Salganik, P. S. Dodds, and D. J. Watts, "Experimental Study of Inequality and Unpredictability in an Artificial Cultural Market," *Science* 311 (2006).

The computer power needed to create one bitcoin

Nathaniel Popper, "There Is Nothing Virtual About Bitcoin's Energy Appetite," *New York Times*, January 21, 2018.

49.

The CEO of a typical company in 1960 made about 20 times as much as its average worker

David Leonhardt, "When the Rich Said No to Getting Richer," *New York Times*, September 5, 2017.

"holding a wolf by the ear"

Thomas Jefferson, letter to John Holmes, April 22, 1820, available at https://www.encyclopediavirginia.org/Letter_from_Thomas_Jefferson_to_John_Holmes_April_22_1820.

It's not the total amount of abundance in the system that promotes goodwill, but the sense that whatever is available is being distributed justly

Robert M. Sapolsky, *Behave: The Biology of Humans at Our Best and Worst* (London: Penguin, 2017).

50.

The economy needn't be a war; it can be a commons

David Bollier, *Think Like a Commoner: A Short Introduction to the Life of the Commons* (Gabriola Island, BC: New Society, 2014).

Reciprocal altruists, whether human or ape, reward those who cooperate with others and punish those who defect

Ernst Fehr and Urs Fischbacher, "The Nature of Human Altruism," *Nature* 425 (October, 2003).

51.

One economic concept that grew out of the commons was called distributism

G. K. Chesterton, *Three Works on Distributism* (CreateSpace, 2009).

cooperative businesses are giving even established US corporations a run for their money

Brad Tuttle, "WinCo: Meet the Low-Key, Low Cost, Grocery Chain Being Called 'Wal-mart's Worse Nightmare,'" *Time*, August 7, 2013.

"The Opposite of Wal-mart," *Economist*, May 3, 2007.

Bouree Lam, "How REI's Co-Op Retail Model Helps Its Bottom Line," *The Atlantic*, March 21, 2017.

In these "platform cooperatives," participants own the platform they're using

Trebor Scholz and Nathan Schneider, *Ours to Hack and to Own: The Rise of Platform Cooperativism, A New Vision for the Future of Work and a Fairer Internet* (New York: OR Books, 2017).

52.

Luckily, according to this narrative, the automobile provided a safe, relatively clean alternative

Stephen Levitt and Stephen J. Dubner, *Freakonomics: A Rogue Economist Explores the Hidden Side of Everything* (New York: William Morrow, 2005).

The problem with the story is that it's not true

Brandon Keim, "Did Cars Save Our Cities from Horses?" *Nautilus*, November 7, 2013.

*They measure improvement as a function of life expectancy or reduc-
tion in the number of violent deaths*

Steven Pinker, *The Better Angels of Our Nature* (London: Penguin, 2011).

*Capitalism no more reduced violence than automobiles saved us from
manure-filled cities*

Nassim Taleb, "The Pinker Fallacy Simplified," FooledBy
Randomness.com/pinker.pdf.

53.

*Online task systems pay people pennies per task to do things that com-
puters can't yet do*

Eric Limer, "My Brief and Curious Life as a Mechanical Turk,"
Gizmodo, October 28, 2014.

55.

*We shape our technologies at the moment of conception, but from that
point forward they shape us*

John Culkin, "A Schoolman's Guide to Marshall McLuhan,"
Saturday Review, March 18, 1967.

*they make decisions just as racist and prejudicial as the humans whose
decisions they were fed*

Ellora Thadaney Israni, "When an Algorithm Helps Send You to
Prison," *New York Times*, October 26, 2017.

*But the criteria and processes they use are deemed too commercially
sensitive to be revealed*

Ian Sample, "AI Watchdog Needed to Regulate Automated
Decision-Making, Say Experts," *Guardian*, January 27, 2017.

Sandra Wachter, Brent Mittelstadt, and Luciano Floridi, "Why
a Right to Explanation of Automated Decision-Making Does Not
Exist in the General Data Protection Regulation," *SSRN*, January 24,
2017.

some computer scientists are already arguing that AIs should be granted the rights of living beings

Antonio Regalado, "Q&A with Futurist Martine Rothblatt," *MIT Technology Review*, October 20, 2014.

57.

In their view, evolution is less the story of life than of data

Ray Kurzweil, *The Age of Spiritual Machines: When Computers Exceed Human Intelligence* (London: Penguin, 2000).

Either we enhance ourselves with chips, nanotechnology, or genetic engineering

Future of Life Institute, "Beneficial AI 2017," https://futureoflife.org/bai-2017/.

to presume that our reality is itself a computer simulation

Clara Moskowitz, "Are We Living in a Computer Simulation?" *Scientific American*, April 7, 2016.

The famous "Turing test" for computer consciousness

Alan Turing, "Computing Machinery and Intelligence," *Mind* 59, no. 236 (October 1950).

58.

The human mind is not computational

Andrew Smart, *Beyond Zero and One: Machines, Psychedelics and Consciousness* (New York: OR Books, 2009).

consciousness is based on totally noncomputable quantum states in the tiniest structures of the brain

Roger Penrose and Stuart Hameroff, "Consciousness in the universe: A review of the 'Orch OR' theory," *Physics of Life Review* 11, no. 1 (March 2014).

The only way to solve consciousness is through firsthand experience
Merrelyn Emery, "The Current Version of Emery's Open Systems Theory," *Systemic Practice and Action Research* 13, no. 5 (2000).

we know consciousness exists because we know what it feels like
Thomas Nagel, *Moral Questions* (Cambridge, UK: Cambridge University Press, 1991).

59.

There is no in-between state
Andrew Smart, *Beyond Zero and One: Machines, Psychedelics and ⲙ̗‍Ⲁⲗⲗⲣ̇ⲓ̇ⲟ̇ⲛⲓ̇ⲟ̇ⲛ̇ Ⲓ̇ⲛ̇ⲓ̇ⲟ̇ⲛ̇ ⲓ̇ⲛ̇ ⲓ̇ⲛ̇ⲓ̇ⲟ̇ⲛ* (Ⲛ̇ⲟ̇ⲓ̇ⲟ̇, ⳹⳹⳹).*
Dan Sperber and Deirdre Wilson, *Relevance: Communication and Cognition* (Hoboken: Wiley–Blackwell, 1996).

63.

When an artificial figure gets too close to reality
Masahiro Morey, "The Uncanny Valley," *Energy* 7, no. 4 (1970).

Roboticists noticed the effect in the early 1970s
Jeremy Hsu, "Why 'Uncanny Valley' Human Look-Alikes Put Us on Edge," *Scientific American*, April 3, 2012.

64.

The easiest way to break free of simulation is to recognize the charade
Johan Huizinga, *Homo Ludens: A Study of the Play-Element in Culture* (Eastford, CT: Martino Fine Books, 2016).

Stage magic confounds our logical sensibilities
Kembrew McLeod, *Pranksters: Making Mischief in the Modern World* (New York: NYU Press, 2014).

65.

The psychologist answered, "Find the others."
Timothy Leary, "You Aren't Like Them," lecture at UC Berkeley, 1968.

Crosstalk breeds solidarity, and solidarity breeds discontent
Trebor Scholz, Uberworked and Underpaid: How Workers Are Disrupting the Digital Economy (Cambridge, UK: Polity, 2016).

Some cults use lie detectors
Lawrence Wright, Going Clear: Scientology, Hollywood, and the Prison of Belief (New York: Vintage, 2013).

But these technologies are just updated versions of the confessionals
John Cornwell, The Dark Box: A Secret History of Confession (New York: Basic Books, 2014).

66.

Awe helps people act with an increased sense of meaning and purpose
P. K. Piff et al., "Awe, the Small Self, and Prosocial Behavior," Journal of Personality and Social Psychology 108, no. 6 (2015).

67.

Historical time was born
Karen Armstrong, A History of God: The 4,000-Year Quest of Judaism, Christianity and Islam (New York: Random House, 1993).

Time and history are nonexistent, and the individual is living in the constant present
Mircea Eliade, The Myth of the Eternal Return, or Cosmos and History (New York: Pantheon, 1954).

68.

Judaism replaced an embodied experience of the divine with abstract monotheism
Douglas Rushkoff, *Nothing Sacred* (New York: Crown, 2003).

We have fallen from the paradise of timeless grace, live in sin, and pray for redemption
Wilhelm Reich, *The Mass Psychology of Fascism* (New York: Farrar, Straus, and Giroux, 1980).

They called it wettiko
Jack D. Forbes, *Columbus and Other Cannibals* (Brooklyn: Seven Stories, 2011).

69.

in this new spirituality we would be *as gods*
John Brockman, "We Are As Gods and Have to Get Good at It: A Conversation with Stewart Brand," *The Edge*, August 18, 2009.

70.

The Soviet–American citizen diplomacy program
Jeffrey J. Kripal, *Esalen: America and the Religion of No Religion* (Chicago: University of Chicago Press, 2008).

This vision still motivates the development of artificial intelligence
Erik Davis, *Techgnosis: Myth, Magic, and Mysticism in the Age of Information* (Berkeley: North Atlantic, 2015).
Pascal-Emmanuel Gobry, "Peter Thiel and the Cathedral," Patheos.com, June 24, 2014, http://www.patheos.com/blogs/inebriateme/2014/06/peter-thiel-and-the-cathedral/ (accessed January 10, 2018).

71.

One Bay Area startup harvests the blood of young people
Maya Kosoff, "This Anti-aging Start-up Is Charging Thousands of Dollars for Teen Blood," *Vanity Fair*, June 2017.

Then, say the chief scientists at the world's biggest internet companies
Ray Kurzweil, *The Singularity Is Near: When Humans Transcend Biology* (London: Penguin, 2005).
Truls Unholt, "Is Singularity Near?" *TechCrunch*, February 29, 2016.

72.

We need a Reason for what we do
Max Horkheimer, *Eclipse of Reason* (Eastford, CT: Martino Fine Books, 2013).

74.

But agriculture also reduced the biodiversity of the human diet
Robert M. Sapolsky, *Behave: The Biology of Humans at Our Best and Worst* (London: Penguin, 2017).

agriculture created the conditions for "feast and famine"
Richard Heinberg, *Our Renewable Future: Laying the Path for One Hundred Percent Clean Energy* (Washington, DC: Island Press, 2016).

We know now, beyond any doubt, that industrial agriculture gets less food out of the ground
John Ikerd, *Small Farms Are Real Farms* (Greeley, CO: Acres USA, 2008).
Raj Patel, *Stuffed and Starved: The Hidden Battle for the World Food System* (New York: Melville House, 2012).
Vandana Shiva, *Stolen Harvest: The Hijacking of the Global Food Supply* (Lexington: University of Kentucky Press, 2015).

Studies by the United Nations and the World Bank

Steve Drucker, *Altered Genes, Twisted Truth: How the Venture to Genetically Engineer Our Food Has Subverted Science, Corrupted Government, and Systematically Deceived the Public* (Fairfield, IA: Clear River Press, 2015).

the earth will run out of topsoil

Chris Arsenault, "Only 60 Years of Farming Left If Soil Degradation Continues," *Scientific American*, December 5, 2014.

75.

Increasing levels of carbon dioxide lead to sharp declines in cognitive ability

David Wallace-Wells, "The Uninhabitable Earth," *New York Magazine*, July 9, 2017.

When the term was coined by a graduate student in 1978

David Holmgren and Bill Mollison, *Permaculture* (Paris: Equilibres d'aujourd'hui, 1990).

recognizing the bigger, subtle cycles of the seasons and the moon

Rudolf Steiner, *Agriculture Course: The Birth of the Biodynamic Method* (Forest Row, UK: Rudolf Steiner Press, 2004).

ancient practices, biodiversity, and livestock rotation can be augmented

Brian Halwell, *Eat Here: Reclaiming Homegrown Pleasures in a Global Supermarket* (New York: Norton, 2004).

Polyface Farms, http://www.polyfacefarms.com/.

76.

Moderns ignored the advice

Kathryn Schulz, "The Really Big One," *New Yorker*, July 20, 2015.

NASA's climate data may as well be tsunami stones

Scientists have been warning us since the 1970s that our environment is in peril. "The Limits to Growth," a widely derided Club of Rome report from 1972, was the first set of data to show how economic growth could not continue to accelerate without depleting the planet's nonrenewable resources. It turned out to be right. In 1992, the Union of Concerned Scientists and more than 1,700 independent scientists, including a majority of living Nobel laureates, issued the "World Scientists' Warning to Humanity," in which they argued that we were pushing the ecosystem beyond its carrying capacity, gravely threatening everything from the oceans and forests to topsoil and the atmosphere—again, with facts and figures. In 2017, the Alliance of Concerned Scientists revisited that alarming data, and determined we're actually worse off than those dire predictions and likely on the brink of a "mass extinction" event.

Brain studies reveal that we relate to our future self the way we relate to a completely different person

Hal Hershfield, "Future Self-Continuity: How Conceptions of the Future Self Transform Intertemporal Choice," *Annals of the New York Academy of Sciences*, 1235, no. 1 (October 2011).

The future is not a discontinuity or some scenario we plan for

Adam Brock, *Change Here Now: Permaculture Solutions for Personal and Community Transformation* (Berkeley: North Atlantic, 2017).

77.

discoveries often come from those who are most intimately dependent on its processes

Clifford D. Conner, *A People's History of Science: Miners, Midwives, and "Low Mechanicks"* (New York: Nation Books, 2005).

"Nature must be taken by the forelock"

Ibid.

78.

treat different parts and processes of the world as separate and independent

Douglas R. Hofstadter, *Gödel, Escher, Bach: An Eternal Golden Braid* (New York: Basic Books, 1999).

There is no backdrop against which reality happens

Julian Barbour, *The End of Time: The Next Revolution in Physics* (New York: Oxford University Press, 2011).

82.

the myth of the individual

Jacob Burckhardt, *The Civilization of the Renaissance in Italy* (New York: Penguin, 1990). Burckhardt's 1860 classic shows how, before the Renaissance, "Man was conscious of himself only as a member of a race, people, party, family, or corporation—only through some general category."

Dr. Faustus in Marlowe's play is often cited

Clarence Green, "Doctor Faustus: Tragedy of Individualism," *Science and Society* 10, no. 3 (Summer 1946).

84.

We push through the contradiction and find a dynamic sensibility on the other side

Robert Nisbet, *The Quest for Community: A Study in the Ethics of Order and Freedom* (Wilmington, DE: Intercollegiate Studies Institute, 2010).

87.

This is the theory of diplomacy that led to the famous Oslo Accords

J. T. Rogers, *Oslo: A Play* (New York: Theater Communications Group, 2017).

the most productive tool for improving international relations

Nancy Snow, *Information War: American Propaganda, Free Speech, and Opinion Control Since 9/11* (New York: Seven Stories, 2003).

. . . face-to-face engagement leverages our evolved capacity to establish rapport

Arthur C. Brooks, "Empathize with Your Political Foe," *New York Times*, January 21, 2018.

88.

social media platforms have demonstrated their ability to induce people to vote or not vote

Robert M. Bond et al., "A 61-million-person experiment in social influence and political mobilization," *Nature* 489 (September 13, 2012).

human-to-human, I-and-you engagement

Martin Buber, *I and Thou* (New York: Scribner, 1958).

95.

genuinely interdependent communities, in which each of us is needed, offer the highest quality of life

Merrelyn Emery, "The Current Version of Emery's Open Systems Theory," *Systemic Practice and Action Research* 13 (2000).

96.

Our courts, democracy, markets, and science are all characterized by competition

Robert Reich, *The Common Good* (New York: Knopf, 2018).

Transparency is really the only choice, anyway

David Brin, *The Transparent Society* (New York: Perseus, 1998).

98.

we're wired from birth and before to share, bond, learn from, and even heal one another
Dr. Mark Filippi, Somaspace homepage, http://somaspace.org.

We humans are all part of the same collective nervous system
Stephen W. Porges, *The Polyvagal Theory: Neurophysiological Foundations of Emotions, Attachment, Communication, and Self-Regulation* (New York: Norton, 2011).

ALSO BY DOUGLAS RUSHKOFF